Quantum Mechanics of Fundamental Systems 1

Series of the Centro de Estudios Científicos de Santiago

Series Editor: Claudio Teitelboim
Centro de Estudios Científicos de Santiago
Santiago, Chile
and University of Texas at Austin
Austin, Texas, USA

IONIC CHANNELS IN CELLS AND MODEL SYSTEMS
Edited by Ramon Latorre

PHYSICAL PROPERTIES OF BIOLOGICAL MEMBRANES AND THEIR FUNCTIONAL IMPLICATIONS
Edited by Cecilia Hidalgo

PRINCIPLES OF STRING THEORY
Lars Brink and Marc Henneaux

QUANTUM MECHANICS OF FUNDAMENTAL SYSTEMS 1
Edited by Claudio Teitelboim

Quantum Mechanics of Fundamental Systems 1

Edited by

Claudio Teitelboim

Centro de Estudios Científicos de Santiago
Santiago, Chile
and University of Texas at Austin
Austin, Texas

Springer Science+Business Media, LLC

Library of Congress Cataloging in Publication Data

Quantum mechanics of fundamental systems 1.

(Series of the Centro de Estudios Científicos de Santiago)
Proceedings of an international meeting, held at the Centro de Estudios Científicos de Santiago, Dec. 16–20, 1985
Includes bibliographies and index.
1. Quantum field theory—Congresses. I. Teitelboim, Claudio. III. Series.
QC174.45.A1Q3628 1988 530.1′43 87-36056

ISBN 978-1-4899-3730-8 ISBN 978-1-4899-3728-5 (eBook)
DOI 10.1007/978-1-4899-3728-5

Originally published by Plenum Press, New York in 1988.
Softcover reprint of the hardcover 1st edition 1988

Contributors

Stephen L. Adler, The Institute for Advanced Study, Princeton, New Jersey 08540

Carlos Aragone, Departamento de Física, Universidad Simón Bolívar, Caracas 1080-A, Venezuela

Laurent Baulieu, Laboratoire de Physique Théorique et Hautes Energies, Université Pierre et Marie Curie, 75252 Paris Cedex 05, France

Ninoslav Bralić, Facultad de Física, Pontificia Universidad Católica de Chile, Santiago 22, Chile

Mario Castagnino, Instituto de Astronomía y Física del Espacio, 1428 Buenos Aires, Argentina; and Instituto de Física de Rosario, 2000 Rosario, Argentina

Rafael Ferraro, Departamento de Matemáticas, Facultad de Ciencias Exactas y Naturales, Universidad de Buenos Aires, Ciudad de Universitaria, 1428 Buenos Aires, Argentina

David J. Gross, Joseph Henry Laboratories, Princeton University, Princeton, New Jersey 08544

Marc Henneaux, Faculté des Sciences, Université Libre de Bruxelles, B-1050, Bruxelles, Belgium; and Centro de Estudios Científicos de Santiago, Santiago 9, Chile

R. Jackiw, Center for Theoretical Physics, Massachusetts Institute of Technology, Cambridge, Massachusetts 02139

A. Khoudeir, Departamento de Física, Universidad de Los Andes, Mérida, Venezuela

Fernando Lund, Departamento de Física, Facultad de Ciencias Física y Matemáticas, Santiago, Chile

Juan Pablo Paz, Instituto de Astronomía y Física del Espacio, 1428 Buenos Aires, Argentina

J. Stephany, Departamento de Física, Universidad Simón Bolívar, Caracas 1080-A, Venezuela

Claudio Teitelboim, Centro de Estudios Científicos de Santiago, Santiago 9, Chile; and Center for Relativity, The University of Texas at Austin, Austin, Texas 71782

Peter van Nieuwenhuizen, Institute for Theoretical Physics, State University of New York at Stony Brook, Stony Brook, New York 11794

Manuel Villasante, Department of Physics, University of California at Los Angeles, Los Angeles, California 90024

Claudia Yastremiz, Instituto de Astronomía y Física del Espacio, 1428 Buenos Aires, Argentina

Preface

Il capitano generale lagrimò per allegrezza e nominò quel capo: Deseado, perchè l'avevamo già gran tempo desiderato.

Antonio Pigafetta
Il Primo Viaggo in torno al Mondo

I would like to take some poetic license in introducing this volume in a way that seems appropriate for a country, like Chile, that looks to the ocean. I believe it was Heisenberg who compared different times in physics with sailing a ship. He said that most of the time we keep our ships in port, or in the protection of a bay. But on a few occasions we go into the open sea, and those occasions are really the great times in theoretical physics, when everything can change. It does not seem totally unwarranted to hope that we are now entering one of those times.

In that spirit, I would like to mention a wonderful book, which in English would be called something like *Chile, Or a Crazy Geography.* (Benjamín Subercaseaux, *Chile o una Loca Geografía*, Editorial Ercilla, Santiago, Chile, 1940.) In it there is a quotation from the diary of Pigafetta, the navigator of Fernando Magellan, who discovered the strait in the south named after him. It says, "...the Captain General cried with happiness and called the cape Desired, because we had desired it for a long time ...." I believe that we probably all had a similar feeling when the wonderful discoveries discussed in this volume took place. We felt that we had desired them for a long time. We had also very much desired the conference on which this volume is based. We are grateful and honored to have had with us friends who could, to use another phrase from the same book, "imagine maps before they exist."

Claudio Teitelboim

Santiago, Chile

Acknowledgments

This book contains the proceedings of a meeting held at the Centro de Estudios Científicos de Santiago from December 16 to 20, 1985. The meeting brought together participants from Europe, Latin America, and the United States.

I would like to thank my colleague Jorge Zanelli for his precious collaboration in organizing the conference, and María Luisa Valdovinos, whose imaginative and tireless efforts transformed theory into practice. Warm thanks are due to Rafael Rosende for his enthusiastic assistance in the preparation of this volume and also to Yolanda Flores for her help.

The meeting took place under a program sponsored by The Tinker Foundation. We are deeply grateful to the Foundation and especially to its President, Martha Muse, for making it all possible.

Contents

Chapter 1

Quaternionic Quantum Field Theory

Stephen L. Adler

The topic of this chapter is whether the standard quantum mechanics, which was invented in Copenhagen in the late 1920s, is the only type of quantum mechanics, or whether there are more general ones; in particular, I want to discuss the possibilities of a quaternionic quantum field theory.

Let me begin by asking (since I am sure many readers will ask), why try to make a new kind of quantum mechanics? I can give two reasons: one is a mathematical reason, which is that we get a better understanding of a system of postulates if we have more than one realization; and specifically we will get a better understanding of complex quantum mechanics, if we understand which features of the usual complex quantum mechanics are more general than others. The second reason is a physical motivation: the current state of particle theory can be represented pretty much as shown in Fig. 1. At low energies we know well what is going on. There is an effective theory, the $SU(3)$ color theory of the strong interactions and the $SU(2) \times U(1)$ weak-electromagnetic theory, which are both described by complex quantum mechanics, and Einstein gravity, which is described by real classical or semiclassical fields. At high energies, what we believe is that there is some common dynamical machine that gives rise to all of these three low-energy theories as effective field theories. And what we do not know, the big question of high-energy physics, is what this

STEPHEN L. ADLER • The Institute for Advanced Study, Princeton, New Jersey 08540.

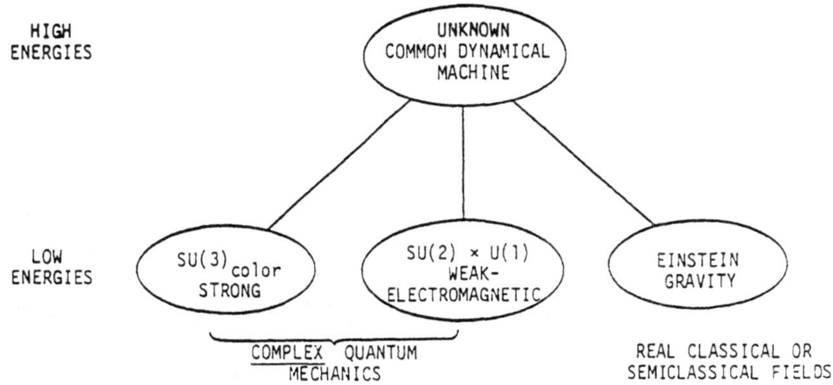

Figure 1. The current state of particle theory.

common dynamical machine is. Attempts over the last fifteen years to construct the common dynamical machine, using local complex quantum mechanics, have run into difficulties. At least, there is no natural, obvious, simple unifying model. So that raises the question whether something new is needed, and I think most of the focus of physics for the last year or two has been on trying new things. For instance, the superstring generalizes from local to nonlocal field theory. Here I shall discuss a generalization in a different direction, which is a generalization from complex numbers to quaternions.

Let me begin by reviewing what it is that distinguishes a quantum system from a classical one. I am going to follow closely here a very nice discussion in Feynman's famous *Reviews of Modern Physics* article, in which he develops the path integral, because in fact essentially everything I do is an attempt to generalize what Feynman does from complex numbers to quaternions.

Let B be a set of attributes that completely specify a state in the quantum mechanical sense, and let P_{ab} be the probability that if a measurement of the set of attributes A gives the result a, then the measurement of B gives b; and let P_{bc}, similarly, be the probability that if the measurement of B gives b, the measurement of C gives c. Classically, the probability that if a measurement of A gives a, then the measurement of C will give c, is simply given by the law of conditional probability. We sum over all intermediate outcomes and just multiply the probabilities. Thus classically, P_{ac} is the sum over b of P_{ab} times P_{bc}:

$$P_{ac} = \sum_b P_{ab} P_{bc}$$

This would be the rule you would use in flipping dice or calculating the outcomes in a card game. (And this, incidentally, is also what is used in the derivation of Bell's theorem that we shall hear about later on.)

However, this law of superposition of probabilities is not generally true of quantum mechanics. Instead, in quantum mechanics, we are guaranteed the existence of a probability amplitude, ϕ_{ab}. (In Dirac notation it is the transition matrix element $\langle b|a\rangle$.) And we are told that the probabilities are the squares in magnitude of the ϕ's: P_{ab} is the magnitude $|\phi_{ab}|^2$, P_{bc} is the magnitude $|\phi_{bc}|^2$, and so forth. And then we are told that ϕ_{ac}, which is the probability amplitude to go from a to c, is equal to the sum over intermediate states b of ϕ_{bc} times ϕ_{ab},

$$\phi_{ac} = \sum_b \phi_{bc}\phi_{ab} = \sum_b \langle c|b\rangle\langle b|a\rangle$$

Now let us notice, for later reference, that when you write things in the Dirac notation, there is a natural factor ordering that comes in also: if you simply write the bra's and ket's in the natural order, from left to right, there is an implicit factor ordering. And that is what will allow for the possibility of going from a commutative to a noncommutative number system, while still preserving the law of superposition of probability amplitudes. To summarize: in classical mechanics, probability is superimposed; in quantum mechanics, probability amplitudes—and not probability—are superimposed.

Let us now ask what kinds of number systems we can use for the ϕ's? There is a theorem, stemming from the work of Birkhoff and von Neumann, which established an axiomatic basis for quantum mechanics, that states that the ϕ's can be real, in which case you have real quantum mechanics; complex, which gives the standard quantum mechanics that we have in all our textbooks; and quaternion, which gives quaternionic quantum mechanics. Let me review briefly what we mean by complex and quaternion numbers. We all know what real numbers are. A complex number, z, has real components

$$z = x + yi$$

where x and y are real, and i^2 is -1. A quaternion number is a generalization of a complex number, with three imaginary units. So a quaternion q has the form

$$q = q_0 + q_1e_1 + q_2e_2 + q_3e_3$$

where the components q_0, q_1, q_2, and q_3 are all real, and e_1, e_2, and e_3 are quaternion units—in the older literature they are often called i, j, and k—and they satisfy an algebra

$$e_ae_b = -\delta_{ab} + \varepsilon_{abc}e_c$$

Writing this out, it says that

$$e_1^2 = e_2^2 = e_3^2 = -1$$

so $e_{1,2,3}$ are all quantities like imaginary units. However, they do not commute;

$$e_1 e_2 = e_3 \quad \text{but} \quad e_2 e_1 = -e_3$$

and similarly for the cyclic permutations, so they obey a noncommutative multiplication law. There is a simple complex matrix realization of the quaternions. We write

$$e_a = -i\tau_a$$

where the τ_a are the standard Pauli matrices, τ_1, τ_2, and τ_3, and this complex matrix realization satisfies the abstract algebra. However, what I will say will not make specific use of the complex realization. I am going to discuss a quantum mechanics where the e's are regarded as fundamental abstract quantities, just as in doing complex number theory we do not normally make use of the fact that i can be given a real matrix representation [i is represented by the matrix: $\begin{pmatrix} 0 & 1 \\ -1 & 0 \end{pmatrix}$, because the square of this matrix is -1].

Now what is it that the real, complex, and quaternion number systems have in common? Let me follow a discussion of Pontryagin and introduce the concept of a number field. It is a number system that has two operations, an addition and a multiplication. The addition and the multiplication are associative, so

$$a + (b + c) = (a + b) + c$$

$$a(bc) = (ab)c$$

and the multiplication is distributive over the addition,

$$a(b + c) = ab + ac$$

The addition is commutative,

$$a + b = b + a$$

but the multiplication is not necessarily commutative, and generally

$$ab \neq ba$$

Finally, there are additive and multiplicative inverses: for every a, there is a $-a$, such that

$$a + (-a) = 0$$

and for every nonzero a, there is an a inverse, such that

$$aa^{-1} = a^{-1}a = 1$$

Let me also introduce the idea of topological structure. A number system is said to be topological, roughly speaking, if continuous variations are possible. Then there is a theorem in Pontryagin that states that the only arcwise-connected, topological number fields are the real, complex, and quaternion numbers. Now, the properties assumed in this theorem are clearly ones that we would like our probability amplitudes in quantum mechanics to have. Note that sacrificing the commutative law of multiplication is not necessarily a problem, because as I showed, if you simply write out the intermediate state sum in Dirac notation, there is already a natural factor ordering. Therefore this permits a natural generalization of quantum mechanics to the case where the matrix elements individually are elements of a noncommutative number system.

Real quantum mechanics was investigated a long time ago by Stueckelberg. What he showed is that to describe physics as we know it, real quantum mechanics must reduce back to complex, or quaternion, quantum mechanics. Basically, what he found is that it must contain an operator, J, such that

$$J^T = -J, \qquad J^2 = -1$$

which commutes with all observables, but that is, of course, just a real representation for i. The reason this is necessary (this is not Stueckelberg's argument, but a simpler one) is that if we want

$$\frac{d}{dt}\int |\Psi|^2 = 0$$

i.e., if we want probability to be conserved, but Ψ to have a nontrivial dynamics, we have to have an imaginary unit present; it is the conjugation from i to $-i$ that makes the Hamiltonian cancel out giving conservation of the wave function normalization. If we do not have the i, then Ψ cannot have a nontrivial dynamics while having a conserved total norm.

So, real quantum mechanics is not interesting; we have to have an imaginary unit. It can be either the imaginary unit of the complex numbers or it can be one of the quaternion imaginary units. So the interesting cases are the standard one, complex quantum mechanics, which describes the observable world, or possibly quaternion quantum mechanics. Now quaternionic quantum mechanics is potentially interesting because Yang–Mills structures start to appear naturally. This was observed first by Yang in a comment at the 1957 Rochester conference that served as the motivation for Finkelstein and collaborators to start looking at the quaternionic case. There followed some early investigations by Finkelstein, Jauch, Speiser, and others, which led to very interesting mathematical structural theorems about quaternion quantum mechanics. But there were stumbling blocks that prevented them from going on to a complete theory. The stumbling blocks were the following:

First, the question of what to use for dynamics. If we simply write

$$\hat{e}\frac{d\Psi}{dt} = H\Psi$$

as a Schrödinger equation, where H is a self-adjoint Hamiltonian and $\hat{e}$ is some quaternion unit, this is simply a rewriting of the complex case, with a different name for the imaginary unit. And although the language of the Finkelstein paper suggests something more general, when they started writing down wave equations, this is essentially what they did. So what they did, in specifics, was actually complex quantum mechanics, giving, in fact, something very close to the Georgi–Glashow model, written in a disguised notation.

The second problem is in dealing with composite systems. If Ψ_a and Ψ_b are quaternion-valued wave functions, then clearly

$$\Psi_a \Psi_b \neq \Psi_b \Psi_a$$

Therefore, we do not have a commutative tensor product. Having a commutative tensor product is the basis for the whole second quantization treatment of field theory. So the second quantization approach to treating many degrees of freedom breaks down in the quaternionic case.

In the rest of this chapter, what I will do is to describe, in outline, a new approach that I have been looking at for the last year, to make a quaternionic quantum mechanics using functional integral methods. It turns out to be possible to proceed fairly far in analogy with the complex case. What I shall do first is to review how, from Dirac's observation in the complex case, one gets the functional integral for a transition amplitude

$$\langle b, t | a, 0 \rangle$$

Then I shall show how this generalizes to quaternions, and sketch briefly how by using the functional integral plus a Gaussian integral formula one can derive the Schrödinger equation—this generalizes what is in Feynman's famous paper; and conversely the generalization of what is in all the standard field theory textbooks—how by starting from the Schrödinger equation, or the equivalent transformation theory, plus the same Gaussian integral formula, one gets back to the functional integral. Thus, all of this standard complex quantum mechanics does generalize to the quaternionic case. However, the analogy is only partial. At the end of the chapter, I shall state a long list of complex quantum theory results that fail in the quaternionic case. Only a subset of complex quantum theory generalizes to the quaternion number system.

The fundamental technical tool on which the whole analysis is based is a Gaussian integral formula. Gaussian integral formulas have a long history in physics, because the theory of wave optics, which is what quantum mechanics originated from, is based on Gaussian integrals. This is where my investigation began. I came across some articles by Dyson showing how to construct quaternion determinants, and reasoned that if there are quaternion determinants then there must be quaternion Gaussian integrals.

Let me begin by reviewing the Gaussian integral formulas in the complex case. This is usually section one in the field theory textbooks: before you develop the path integral, you first state the Gaussian integral formula, because that is the essential technical tool in developing the standard Feynman path integral. So let me first state it for bosons. Let z and u be column vectors containing N complex numbers, and let A be an $N \times N$ complex matrix. Then the standard Gaussian integral formula for complex numbers says that

$$\left(\prod_1^N \int \frac{d\bar{z}\, dz}{2\pi i}\right) e^{-\bar{z}Az+\bar{u}z+\bar{z}u} = (\det A)^{-1} e^{\bar{u}A^{-1}u}$$

This is very simply derived; we shift the integration variable z by something proportional to A^{-1}, which eliminates the linear terms, so we have completed the square. The $\exp(\bar{u}A^{-1}u)$ piece factors out, and then we simply make a unitary transformation to diagonalize A, and integrate mode by mode, leading to a product of eigenvalues in the denominator that gives determinant A. That is a sketch of the proof; a neater derivation can of course be given.

We shall also need a similar formula for fermions. Now in field theory, to handle fermions one has to introduce peculiar objects called Grassmann numbers. They are anticommuting numbers. In other words,

$$\psi_a \psi_b + \psi_b \psi_a = 0$$

for any a and b. That means that if a and b are equal, we have

$$\psi_a^2 = 0$$

So the square of a Grassmann number is 0, and any two Grassmann numbers anticommute. Then we define integration of a Grassmann variable by the formulas

$$\int d\psi_a \psi_b = \delta_{ab}$$

$$\int d\psi_a(\text{any other monomial}) = 0$$

This means that $\int d\psi_a$ is effectively simply a derivative with respect to the fermionic variable ψ_a at $\psi_a = 0$. But these derivatives anticommute, because the Grassmann numbers themselves anticommute.

Now let ψ, ξ be column vectors of N complex Grassmann numbers. The analog of the bosonic formula—and this is also in the first section of any standard field theory textbook—says

$$\left(\prod_{1}^{N}\int d\bar{\psi}\,d\psi\right)e^{-\bar{\psi}A\psi+\bar{\xi}\psi+\bar{\psi}\xi}=\det Ae^{\bar{\xi}A^{-1}\xi}$$

A is, again, a general $N \times N$ complex matrix. In the complex case, I do not have to restrict A to be Hermitian or anti-Hermitian, but in the quaternionic case I will have to say something more specific about what A is. The reason det A appears upstairs in the fermionic formula is the antisymmetry of the Grassmann multiplication: we simply Taylor-expand the formula, and it terminates at the nth-order term, because there we have exhausted all the $\int d\bar{\psi}$'s and $\int d\psi$'s, and we just get the expansion for the determinant as a sum over products of one element in each row and column with an alternating sign.

In order to make a quaternionic analog of these Gaussian integrals, we replace the complex variable, z, by a quaternion ϕ, which is

$$\phi = \phi_0 + \phi_1 e_1 + \phi_2 e_2 + \phi_3 e_3$$

and which has a quaternionic conjugate

$$\bar{\phi} = \phi_0 - \phi_1 e_1 - \phi_2 e_2 - \phi_3 e_3$$

obtained by simply reversing the signs of all three quaternion units. That is an analog of complex conjugation. In the above equation, the ϕ_a's are real. And the reason they are real is that complex quaternions do not form a division algebra because

$$(1 + ie_3) \times (1 - ie_3) = 0$$

so we can have nonzero elements of the algebra multiplying to give zero. However, the Birkhoff-von Neumann construction says that one really wants to have a division algebra. A simple heuristic way of seeing this is, if one wants to get propagators, everything in one's algebra of scalars must have an inverse, so it is not possible to form propagators without a division algebra. Therefore we must deal with the quaternions over the real numbers. The ψ's are similarly Grassmann quaternions. In other words, ψ is a quaternion whose coefficients are Grassmann elements of just the form I discussed before,

$$\psi = \psi_0 + \psi_1 e_1 + \psi_2 e_2 + \psi_3 e_3$$

$$\bar{\psi} = \psi_0 - \psi_1 e_1 - \psi_2 e_2 - \psi_3 e_3$$

Now if we want to do real-time quantum mechanics, and develop the quaternion analog of the familiar complex case, the analog of the complex

e^{iL} will be $e^{\tilde{L}}$, where $\tilde{L}$ is a quaternion imaginary Lagrangian. $\tilde{L}$ will be

$$\tilde{L} = e_1 L_1 + e_2 L_2 + e_3 L_3$$

with $L_{1,2,3}$ real. So to do the analog of the complex, real-time path integral, we need to consider oscillatory quaternion Gaussian integrals. To evaluate such integrals we need an integration measure. It turns out that the natural integration measure that replaces $d\bar{z}\,dz$ is $d\phi$, where $d\phi$ is simply the product of the differentials of the real components,

$$\int d\bar{z}\,dz \to \int d\phi \equiv \int d\phi_0\, d\phi_1\, d\phi_2\, d\phi_3$$

and similarly, the analog of $d\bar{\psi}\,d\psi$ is

$$\int d\bar{\psi}\,d\psi \to \int d\psi \equiv \int d\psi_0\, d\psi_1\, d\psi_2\, d\psi_3$$

The reason that this is the right measure is that we can show—and this requires some work—that the measure defined this way is invariant under arbitrary quaternion unitary transformations.

From here on things get more complicated, and I shall only state a result: Because the e_a's, the quaternion units, do not commute, when we complete the square we find that we cannot then factor the source term out of the integral, because when a and b are not commuting operators,

$$e^{a+b} \neq e^a e^b$$

So we find that the source dependence is not exponential in general, when we figure out how to do the integrals. It is not just an analog of the $e^{\bar{u}A^{-1}u}$ or $e^{\bar{\xi}A^{-1}\xi}$ of the complex case. But now comes the first surprise. When we perform the integrations, what we find is that when the number of bosonic integration variables is equal to the number of fermionic integration variables, the sources do reexponentiate, and we get the following formula:

$$\lim_{\varepsilon\to 0}\left(\prod_{i=1}^{M}\int d\phi^i\,d\psi^i\right)(4\pi^2)^{-M}\exp(-\bar{\phi}A\phi - \bar{\psi}B\psi + \bar{u}\phi - \bar{\phi}u + \bar{\xi}\psi$$
$$+ \bar{\psi}\xi + \mathcal{K} - \varepsilon\bar{\phi}\phi) = \det^2 B\,\det^{-1}(A^{\dagger}A)\exp(-\bar{u}A^{-1}u + \bar{\xi}B^{-1}\xi + \mathcal{K})$$

Here $\mathcal{K}$ is a quaternion-imaginary constant, and apart from the infinitesimal convergence term $\varepsilon\bar{\phi}\phi$, everything in the exponents is quaternion imaginary, provided one makes the choice that A is a quaternion anti-Hermitian matrix, and B is a quaternion Hermitian matrix. So we find the surprising feature that the formulas that are valid for bosons and fermions separately in the complex case are valid for quaternions only when the bosons and fermions are combined in equal numbers. I do not know the deep reason for this;

the way I derived this result was by getting a recursion relation that describes what happens as I add one integral at a time. And then, although the computations look very different, it turns out that the recursion relations for bosons and fermions are precisely inverse to each other.

Now let us go back to complex quantum mechanics and sketch how Dirac and Feynman got the path integral. Let me write down a time line for a system with a single coordinate, x, and divide the segment of the time line from t to $t + T$ into N little intervals of width Δt:

t_N	t_{j+1}	t_j	t_0	
⋮	⋮	⋮	⋮	Time line
$t + T$			t	

$$t_{j+1} - t_j = \Delta t = T/N$$

Now, Dirac, in the 1930's, made the famous observation that the transformation function from time t_j to time t_{j+1}, in the limit as Δt goes to 0, is simply related to the classical Lagrangian,*

$$\langle x_{j+1}, t_{j+1} | x_j, t_j \rangle = (1/\zeta)\, e^{i\Delta t L(x_{j+1/2}, \dot{x}_{j+1/2}, t_{j+1/2})}$$

Here $L(x, \dot{x}, t)$ is a classical Lagrangian, and $x_{j+1/2}$, $\dot{x}_{j+1/2}$, and $t_{j+1/2}$ are just the position, the velocity, and the time evaluated by the trapezoidal rule at the midpoint of the interval,

$$x_{j+1/2} = \tfrac{1}{2}(x_{j+1} + x_j)$$

$$\dot{x}_{j+1/2} = (x_{j+1} - x_j)/\Delta t$$

$$t_{j+1/2} = t_j + \Delta t/2$$

What Feynman did to make the path integral—beyond what is partly implicit in Dirac's paper—was to show that one could derive the Schrödinger equation from it, and essentially erect the whole quantum mechanics apparatus from the path integral. Feynman put together many infinitesimal time factors; for each little infinitesimal interval, we know the transformation function: it is just $e^{i\Delta t L}$, with an error that is of order $(\Delta t)^2$. That allows us to do a Riemann sum, and so therefore the finite time transformation function, $\langle x', t + T | x, t \rangle$, is just the product over all the intermediate time slices, $j = 1$ to $N - 1$, of integrals of the infinitesimal transformation function,

$$\langle x', t + T | x, t \rangle = \left(\prod_{j=1}^{N-1} \int dx_j \right) \zeta^{-1} e^{i\Delta t L(N-1/2)} \zeta^{-1} e^{i\Delta t L(N-3/2)} \cdots \zeta^{-1} e^{i\Delta t L(1/2)}$$

And in the limit $N \to \infty$, this gives the Feynman path integral.

* Note: $L(x_{j+1/2}, \dot{x}_{j+1/2}, t_{j+1/2})$ is abbreviated below to $L(j + 1/2)$.

This construction of quantum mechanics invites a quaternionic analog. Let us simply take as our postulate that the transformation function from a quaternionic state, where the coordinates ϕ and ψ have values ϕ_j and ψ_j at time t_j, to one where they have values ϕ_{j+1}, ψ_{j+1} at time t_{j+1}, is equal to some constant, times the exponential of Δt times something which we shall call my quaternion Lagrangian,

$$\langle\{\phi_{j+1}, \psi_{j+1}\}, t_{j+1}|\{\phi_j, \psi_j\}, t_j\rangle$$
$$= \zeta^{-1} \exp[\Delta t\tilde{L}(\{\phi_{j+1/2}, \dot{\phi}_{j+1/2}, \psi_{j+1/2}, \dot{\psi}_{j+1/2}\}, t_{j+1/2})]$$

But now $\tilde{L}$ has to be quaternion imaginary, so it will be a sum of three terms,

$$\tilde{L} = L_1 e_1 + L_2 e_2 + L_3 e_3$$

with $L_{1,2,3}$ real. Thus $\tilde{L}$ will be some function that specifies the dynamics—and again, I shall evaluate it at the midpoint of the interval. But again, just by combining infinitesimal time steps, I can get a finite time formula that says that the transformation function from t_0 and some initial state at time t_0 to t_N and some final state at time t_N is just the product over all intermediate states, integrating over all the intermediate state variables, times the infinitesimal transformation functions,

$$\langle\{\phi_N, \psi_N\}, t_N|\{\phi_0, \psi_0\}, t_0\rangle$$
$$= \left[\prod_{j=1}^{N-1}\left(\prod_{i=1}^{M}\int d\phi_j^i\, d\psi_j^i\right)\right] \zeta^{-1} e^{\Delta t\tilde{L}(N-1/2)} \zeta^{-1} e^{\Delta t\tilde{L}(N-3/2)} \cdots \zeta^{-1} e^{\Delta t\tilde{L}(1/2)}$$

In taking the product now the ordering is important, because the infinitesimal transformation functions do not commute, but we saw that the Dirac notation gives us a natural factor ordering, and in the limit $N \to \infty$ we get what physicists call a time-ordered product. This recipe automatically satisfies the principle of probability amplitude superposition, because if I break the product at some intermediate time, t_I, the product of infinitesimal factors from t_0 to t_I is just a transformation function from t_0 to t_I, there is an integration over the variables at time t_I, and then the remaining string of infinitesimal factors is a transformation function from t_I to t_N,

$$\langle\{\phi_N, \psi_N\}, t_N|\{\phi_0, \psi_0\}, t_0\rangle$$
$$= \left(\prod_{i=1}^{M}\int d\phi_I^i\, d\psi_I^i\right)\langle\{\phi_N, \psi_N\}, t_N|\{\phi_I, \psi_I\}, t_I\rangle\langle\{\phi_I, \psi_I\}, t_I|\{\phi_0, \psi_0\}, t_0\rangle$$

So the quaternionic generalization of the path integral satisfies the principle of superposition of probability amplitudes, which, as I stated at the beginning, I believe to be the fundamental statement of what we mean by quantum mechanics.

We are ready now to take the next step. We have constructed a path integral; the question now is, can we derive a Schrödinger equation from the path integral; and vice versa: from that Schrödinger equation, can we go back and mimic the standard textbook derivation and rederive the path integral? Let us first recall that in the complex case, to derive the Schrödinger equation one has to specialize the kinetic term by assuming it is a quadratic form in time derivatives. We write

$$L = \tfrac{1}{2}\dot{x}^2 - V(x)$$

and then from this Feynman derived the Schrödinger equation. Similarly, in the quaternionic case, if we make an analogous specialization that is suggested by the structure of the Gaussian integral formula, we can derive a Schrödinger equation. The specialization is to write $\tilde{L}$ as

$$\tilde{L} = \tilde{L}_{\text{kin}} - \tilde{V}(\{\phi, \psi\}, t)$$

where $\tilde{V}$ is a completely arbitrary quaternion imaginary potential. So I have interactions between my modes. $\tilde{L}_{\text{kin}}$ is a quadratic form

$$\tilde{L}_{\text{kin}} = \sum_{i=1}^{M} (\tfrac{1}{2}\dot{\bar{\phi}}^i e_3 \dot{\phi}^i + \tfrac{1}{2}\dot{\bar{\psi}}^i \dot{\psi}^i)$$

which is suggested by simply using my Gaussian integral formula, taking the Hermitian matrix B to be 1, and taking the anti-Hermitian matrix A to be any arbitrary quaternion unit, say e_3. The boson structure is essentially conventional, while the fermion structure is unconventional because it has a second-order wave operator (and this implies an indefinite metric Hilbert space, when we look in detail at the fermion state structure). But it seems to be necessary to use this second-order fermion wave operator in order to get the construction to work. That could be a problem when we try to apply it to physics, because what we see are first-order fermions. If the quaternionic path integral is to describe physics, there will have to be some mechanism for splitting half the states off, and that is something I do not yet know how to do. The e_3 in the first term is basically arbitrary, since under the quaternion gauge transformation of replacing ϕ by $q\phi$, where q is a unit quaternion with $\bar{q}q = 1$, e_3 gets replaced by $\bar{q}e_3 q$, and that allows us to rotate e_3 to any other imaginary quaternion direction. That is, I need the quaternion imaginary unit there to get an imaginary $\tilde{L}$, but the choice is completely arbitrary and can be simply rotated by a change in variable, and this change of variable is an invariance of the integration measure.

Having specialized $\tilde{L}$ as described, now we find our second surprise: Using only the Gaussian integral formula, one can derive the Schrödinger equation from the functional integral, and vice versa, in the quaternionic case. To derive the Schrödinger equation, for example, we define a wave function, Ψ, of the coordinates at time t, as

$$\Psi(\{\phi, \psi\}, t) = \langle\{\phi, \psi\}, t|\psi\rangle$$

and then we write

$$\Psi(\{\phi, \psi\}, t + \Delta t) = \left(\prod_i \int d\phi_0^i \, d\psi_0^i\right) \zeta^{-1} e^{\Delta t \tilde{L}(1/2)} \Psi(\{\phi_0, \psi_0\}, t)$$

and Taylor-expand in Δt. This is the standard Feynman derivation. And what we find is that the Feynman derivation goes through when we use the Gaussian integral formula. The reason is that the Gaussian formula, when there are equal numbers of fermions and bosons, allows us to integrate out all the modes when we are talking about the potential energy, but the potential still reexponentiates. Or we can integrate out all the modes save one, when we are considering the kinetic energy term, and the kinetic energy of the one mode we are considering reexponentiates, so that we get a universal structure independent of how many modes there are. This reexponentiation feature is essential for the derivation to go through, and what we find is that we get a Schrödinger equation whose form is totally independent of the number of modes. In the process, the normalization constant, ζ, is fixed to be $\zeta = (4\pi^2)^M$. Because of the use of a second-order fermion wave operator, there is no Δt in the path integral normalization; it is purely a phase-space factor. The Δt's from the bosons cancel the Δt's from the fermions, and so we get something that looks mathematically better behaved than we do in the standard path integral.

The Schrödinger equation that we get has the form

$$\frac{d\Psi}{dt} = -\tilde{H}\Psi$$

where $\tilde{H}$ is the quaternion imaginary Hamiltonian—it is the analog of iH in the complex case. $\tilde{H}$ contains a kinetic term plus a potential term, and the kinetic term has the form

$$\tilde{H}_{\text{kin}} = -\frac{1}{6} \sum_{i=1}^{M} (\bar{D}_{\phi^i} e_3 D_{\phi^i} + \bar{D}_{\psi^i} D_{\psi^i})$$

where D_ϕ and D_ψ are derivative operators that are constructed in a way analogous to the construction of ϕ and ψ. The one peculiar feature is that in the complex analog of this formula, instead of getting $\frac{1}{6}$ as the coefficient in the kinetic term, we get $\frac{1}{2}$. The extra factor of 3 comes from averaging over quaternionic rotations, and it came as a surprise. That was why I wanted to be very careful to check that I could do the inverse derivation and get back to my original Lagrangian formula, which does have a coefficient of $\frac{1}{2}$ in it, to make sure that the $\frac{1}{6}$ was not a mistake.

Now the converse involves proving that

$$\langle\{\phi, \psi\}, t|e^{-\Delta t\tilde{H}}|\{\phi_0, \psi_0\}, t\rangle \leftrightarrow (4\pi^2)^{-M} e^{\Delta t(\tilde{L}_{\rm kin}-\tilde{V})}$$

where "$\leftrightarrow$" indicates equivalence inside the functional integral. And, in fact, a complicated derivation shows that we can rederive the functional integral from the transformation theory, again using just the Gaussian integral formula as a tool.

Let me now show how in one respect, the standard quantum mechanics familiar from the complex case breaks down in the quaternionic generalization. In the quaternionic case, we do not have a correspondence principle leading to quaternionic classical equations as a correspondence limit. There are many ways of seeing this—one is to show that the Ehrenfest theorems break down—but let me show that if we attempt to do a stationary phase approximation, we do not get quaternionic classical equations as a stationary phase approximation. There are two reasons for that. One is that those infinitesimal factors, $\Delta t\tilde{L}$, are noncommuting. Therefore, their product is not simply the Riemann sum of the $\Delta t\tilde{L}$'s,

$$\prod e^{\Delta t\tilde{L}} \neq e^{\sum \Delta t\tilde{L}}$$

Suppose that we ignore this, and say that in some approximation we can neglect the path ordering, and call the sum of $\Delta t\tilde{L}$'s the classical quaternion action, $\tilde{S}$, and ask: What happens if we require that $\tilde{S}$ be stationary? In that case we get three variational principles, because the three components of $\tilde{S}$, along the three quaternionic directions, individually have to be stationary,

$$\delta\tilde{S} = \delta\sum \Delta t\tilde{L} = 0$$

$$\Rightarrow \delta S_1 = \delta S_2 = \delta S_3 = 0$$

and three variational principles are not in general stationary on the same orbits. So if I write down a general quaternion imaginary classical action, I cannot in general demand that it be stationary, because I get three different sets of Euler equations. I can write down special classical actions, where the stationary orbits of the three components are the same, but in general it is not true. On the other hand, it makes perfect sense to use these $\tilde{L}$'s as infinitesimal phase factors, even if I cannot require that there be corresponding classical orbits. Hence the correspondence limit of quaternionic quantum mechanics is not a quaternionic classical mechanics. If there is a correspondence principle, which I believe there is, it is that quaternionic quantum mechanics will map into complex quantum mechanics, and then complex quantum mechanics will have a correspondence principle of the usual sort.

Let me then come to a conclusion. One can construct a quaternionic quantum mechanics with arbitrary numbers of degrees of freedom, and in particular this means one can take the limit of M going to infinity—which is field theory—and so there is no difficulty in this construction in making a quaternionic quantum field theory. The construction depends on having complete boson-fermion symmetry, because the Gaussian integral formula has an especially simple form when the numbers of bosons and fermions are equal. One has a Schrödinger equation, a Dirac transformation theory, and a functional integral. On the other hand, a lot of the apparatus of complex quantum mechanics is absent. You do not have a commuting tensor product, asymptotic states, or an S matrix, except in a complex specialization. You do not have a canonical formalism, coherent states or a Euclidean continuation. All of these nice things that we associate with complex quantum mechanics are present only if the quaternionic theory is first specialized to a complex one. So the conclusion is that there seems to be a new kind of quantum mechanics. There are many interesting formal questions to study, and we hope that the analysis of these questions will eventually allow us to settle whether it is relevant for particle physics.

SUGGESTED READINGS

Some of the introductory material in this chapter was similar to a talk given at the Bohr Symposium and reprinted here with the permission of Gordon and Breach; see also the following reference:

S. L. Adler, in F. A. Harris, S. Pakvasa, and S. F. Tuan (eds.), *Proceedings of the Tenth Hawaii Conference on High Energy Physics*, University of Hawaii Press, Manoa/Honalulu, 1986, p. 481.

The more technical material discussed can be found in the following articles:

S. L. Adler, *Commun. Math. Phys. 104*, 611 (1986). Details of the Gaussian integral formula and applications to quaternionic quantum mechanics.

S. L. Adler, in Progress in Electroweak Interactions, *Proceedings of the Recontres de Moriond, 1986*, J. Tran ThanhVan (ed.), Editiones Frontieres, Gif Sur Yvette, 1986. Speculations about connections with quark-lepton substructure.

For related work done after the Chile Summer Meeting, see the following:

S. L. Adler, *Phys. Rev. Lett. 57*, 167 (1986); *Phys. Rev. D 34*, 1871 (1986). Study of a possible connection between quaternionic quantum mechanics and CP nonconservation.

Chapter 2

Dynamics of Self-Dual Massive Gravity

Carlos Aragone and A. Khoudeir

1. INTRODUCTION

In three dimensions it has been shown [1] that the self-dual massive vector model [2] is equivalent to topologically massive electrodynamics [3] through the discovery of the associated first-order master action. This equivalence holds even in the presence of external sources.

In this talk we present some new results [4] concerning the existence of a massive self-dual spin-2 theory that appears naturally associated with topological massive gravity [5]. Some conceptual differences emerge due to the third-order character of this theory, as compared with the second-order character of topologically massive electrodynamics. The self-dual theory, however, continues to be a first-order model, as we shall see.

The first-order self-dual massive spin-2 field equation will be easily found after writing topologically massive gravity in an intrinsically geometric form language and then linearizing it.

Gravity is carried by the dreibein 1-forms $e^a \equiv dx^m e^a_m$, where $a \in (0, 1, 2)$ is a local Lorentz index and $m \in (0, 1, 2)$ is the standard world index. The Lorentz affinities can be packed into three additional 1-forms

Carlos Aragone • Department of Physics, Simón Bolívar University, Caracas 1080-A, Venezuela. A. Khoudeir • Department of Physics, University of the Andes, Merida, Venezuela.

$\omega^b = dx^m \omega^b_m$, where one is taking advantage that in three dimensions it is simpler to use ω^b_r than the usual antisymmetric components $\omega_{mab} \equiv -\varepsilon_{abc}\omega^c_m$. ε^{abc} is defined by $\varepsilon^{012} \equiv +1$; the metric η^{ab} has the signature $(-,+,+)$.

The fundamental second-order objects are the three 2-forms

$$R^{*a} \equiv d\omega^a + (\tfrac{1}{2})\varepsilon^a_{bc}\omega^b\omega^c$$
$$\equiv (\tfrac{1}{2})dx^{mn}(\partial_m\omega^a_n, -\partial_n\omega^a_m + \varepsilon^a_{bc}\omega^b_m\omega^c_n) \tag{1a}$$

which after another dualization leads to the double dual of the Riemann tensor

$$R^{**pa} \equiv (\tfrac{1}{2})\varepsilon^{pmn}R^{*a}_{mn} \equiv (\tfrac{1}{4})\varepsilon^{pmn}\varepsilon^{abc}R_{mnbc}$$
$$= e[R^{ap} - (\tfrac{1}{2})e^{pa}R] \equiv eG^{ap} \tag{1b}$$

where $e \equiv \det e^a_m$.

The Ricci tensor $R_{ma} \equiv e^{nb}R_{mnab}$ can be seen to have the value

$$R^{am} \equiv e^{pa}e^{mb}R_{pb} = e^{-1}(R^{**ma} - e^{ma}R^{**}) \tag{1c}$$

in terms of the double dual. The last set of elements we want to introduce are the 1-forms $E^a \equiv dx^m E^a_m$ defined to be

$$E^a_n = R^a_n - (\tfrac{1}{4})e^a_n R = e^{-1}[R^{**a}_n - (\tfrac{1}{2})e^a_b R^{**}] \tag{1d}$$

We assume vanishing torsion for this pure bosonic system. That means that the associated 2-forms $2S^a \equiv de^a + \varepsilon^a_{bc}\omega^b e^c$ must vanish, providing for the Lorentz affinity the value

$$e^r_a e^p_b \omega^b_r \equiv \omega^p_a = -e^{-1}\varepsilon^{pmn}\partial_m e_{na} + (2e)^{-1}e^p_a\varepsilon^{mnb}\partial_m e_{nb} \tag{2}$$

Topologically massive gravity has been shown [5] to have the third-order action

$$I_\mu \equiv 2\kappa^{-2}\langle e_{pc}R^{**pc}(\omega)\rangle + \mu^{-1}\kappa^{-2}\langle \omega_{ma}R^{**ma}(\omega)$$
$$-\tfrac{1}{6}\varepsilon^{mnp}\varepsilon^{abc}\omega_{ma}\omega_{nb}\omega_{pc}\rangle \tag{3}$$

Independent variations of the dreibein field e_{pc} yield the equation that determines the evolution of the system,

$$R^{*a} - \mu^{-1}(dE^a + \varepsilon^a_{bc}\omega^b E^c) = 0 \tag{4}$$

On flat space where $e^a_p = \delta^a_p + \kappa h^a_p$ only linear terms are kept alive. These equations become

$$d\omega^a_L - \mu^{-1}\, dE^a_L = 0 \tag{5a}$$

They are locally equivalent to $\omega^a_L - \mu^{-1}E^a_L = d\chi^a$, where χ^a are three arbitrary scalar functions. We set them equal to zero getting a second-order system of field equations

$$\omega^a_L - \mu^{-1}E^a_L = 0 \tag{5b}$$

From equations (2) and (1d) we obtain the values of ω^{p}_{aL} and E^{a}_{pL}

$$\omega^{p}_{aL} = -\varepsilon^{pmn}\partial_m h_{na} + (\tfrac{1}{2})\delta^{p}_{a}\varepsilon^{qmn}\partial_m h_{nq} \tag{6a}$$

$$E^{a}_{pL} = \varepsilon^{amn}\partial_n \omega_{np} - (\tfrac{1}{2})\delta^{a}_{p}\varepsilon^{bmn}\partial_m \omega_{nb} \tag{6b}$$

After introducing these values into equation (5b) we get (understanding that ω^{p}_{aL} becomes ω^{p}_{a})

$$-\varepsilon^{amn}\partial_m h_{np} = \mu^{-1}\varepsilon^{amn}\partial_m \omega_{np} \tag{5c}$$

which again can be, locally, integrated out leading to $-h_{np} = \mu^{-1}(\omega_{np} + \partial_n l_p)$, with arbitrary parameters l_p. We again set these to zero arriving at the first-order self-dual field equations:

$$\begin{aligned} -h_{np} &= \mu^{-1}\omega_{np}(h) \\ &= \mu^{-1}(-\varepsilon^{rs}_{p}\partial_r h_{sn} + (\tfrac{1}{2})\eta_{np}\varepsilon^{rsq}\partial_r h_{sq}) \end{aligned} \tag{7a}$$

which are equivalent to

$$\mu^{-1}\varepsilon^{rs}_{p}\partial_r h_{sn} = h_{np} - \eta_{np}h_{rr} \tag{7b}$$

Since (5a) is linearized topologically massive gravity, we know its dynamics content [4]: it propogates only one degree of freedom, corresponding to the massive excitation carrying spin +2 or −2. We shall see that both the second-order system (5b) and (6a) and (6b) and the self-dual one (7b) have the same dynamical content: they propagate only one degree of freedom, which corresponds to a pure massive spin +2 or −2 physical excitation.

2. ACTIONS

The self-dual action for massive spin 2 is

$$I_{\rm SD} \equiv (2\mu)^{-1}\langle w_{pm}\varepsilon^{prs}\partial_r w_{sm}\rangle - 2^{-1}\langle w_{mp}w_{pm} - w_{rr}{}^2\rangle \tag{7c}$$

It is a straightforward generalization of the vector self-dual action of Ref. 2. Independent variations of w_{pm} (which is not symmetric) give the field equations (7b).

The second-order system (5b) and (6a) and (6b) can be deduced by making independent variations of $(h_{pa}; \omega_{rb})$ in the first-order intermediate master action:

$$\begin{aligned} I_{\rm interm} \equiv &-(\tfrac{1}{2})\langle \omega_{np}\omega_{pn} - \omega_{nn}\omega_{pp}\rangle - (\tfrac{1}{2})\mu\langle h_{pa}\varepsilon^{pmn}\partial_m h_{na}\rangle \\ &-\langle \omega_{pa}\varepsilon^{pmn}\partial_m h_{na}\rangle \end{aligned} \tag{8}$$

It can be checked that independent variations of ω_{pa} and h_{rb} lead to the system (6a) and (5c). This last equation in three dimensions is locally equivalent to $h_{pa} = -\mu^{-1}\omega_{pa}$, which, if inserted into the above action (8), reproduces our self-dual action (7c). Then they are dynamically equivalent.

There are also is a master action for the third-order system (5a). It has the form

$$I_M(h, \omega, E) \equiv \langle h_{pa}\varepsilon^{pmn}\partial_m E_{na}\rangle - \mu\langle h_{pa}\varepsilon^{pmn}\partial_m\omega_{na}\rangle$$
$$+ \langle E_{np}\omega_{pn} - E_{nn}\omega_{pp}\rangle - \langle \omega_{pa}\varepsilon^{pmn}\partial_m\omega_{na}\rangle \tag{9a}$$

Independent variations of h_{pa} yield equation (5a). Variations of E_{pr} give equation (8a) in the form

$$\varepsilon^{pmn}\partial_m h_{nr} + \omega_{rp} - \eta_{rp}\omega_{ss} = 0 \tag{9b}$$

while variations with respect to the affinity ω_{pr} lead to

$$2\varepsilon^{pmn}\partial_m\omega_{nr} = E_{rp} - \eta_{rp}E_{ss} - \mu\varepsilon^{pmn}\partial_m h_{nr} \tag{9c}$$

If one adds to this equation μ times (9b) and takes into account (5a) in its form (5b) one arrives at

$$E_{rp} - \eta_{rp}E_{ss} = \varepsilon_p^{\ mn}\partial_m\omega_{nr} \tag{9d}$$

The set of equations (5a), (9b), and (9d) constitutes the first-order form of linearized topologically massive gravity. If one introduces the value of ω_{ra} provided by (5b) into I_M one obtains that also E_{pa} and consequently ω_{pa} obey the self-dual equations (7b) since

$$I_M(h, \omega = \mu^{-1}E, E) = 2\mu^{-1}I_{\rm SD}(E) \tag{10}$$

On the other hand, insertion of $\omega = \omega(\partial h)$ and of $E = E(\partial\omega)$ into I_M will lead to the third-order action of (linearized) topologically massive gravity. It is worth observing that while $I_{\rm SD}$ is not invariant under the natural gauge transformations ($\sim\partial_m\zeta_p$), both $I_{\rm interm}$ and I_M are.

3. CLASSICAL DYNAMICS

In order to strengthen and confirm our arguments regarding the physical content of the self-dual massive spin-2 actions, we shall consider both its light-front dynamics and its space-time canonical Hamiltonian formulation.

In three dimensions we take the null coordinates $x^{\pm} \equiv 2^{-1/2}(x^0_{\mp}x^2)$. Times goes along x^+ while x^- is considered a spacelike variable ($\partial f/\partial x^+ \equiv \dot{f}$ and $\partial f/\partial x^- \equiv f'$). One expands $I_{\rm SD}$ as defined in (7c) in terms of $(w_{ab};\ a_{a1};\ w_{1b};\ w_{11})$ $a, b \in (+, -)$ according to the natural light-front $2+1$ splitting of the three-dimensional space-time. We take $\eta_{+-} = -1$ and $\varepsilon^{+1-} =$

+1. We find in a first stage that ω_{+r} constitute three Lagrangian multipliers associated with three differential constraints whose solutions are $[\Delta_\mu \equiv -\partial_1^2 + \mu^2 > 0]$

$$w_{-+} = -(\partial_1 + \mu)(\Delta_\mu)^{-1}w'_{1+} + \mu(\partial_1 + \mu)(\Delta_\mu)^{-1}w_{11} \tag{11a}$$

$$w_{-1} = (\partial_1)^{-1}w'_{11} - \mu(\partial_1)^{-1}w_{1-}, \qquad w_{--} = (\Delta_\mu)^{-1}(-\partial_1 + \mu)w'_{1-} \tag{11b}$$

Insertion of these values for w_{-r} into the initial action (7c) leads to

$$\begin{aligned} I_{\text{SD,red}} = {} & \langle w_{1-}(\partial_1)^{-1}\dot{w}_{11} + w_{1-}(\Delta_\mu)^{-1}(\partial_1 + \mu)\dot{w}_{11}\rangle \\ & - \langle w_{11}\mu(\Delta_\mu)^{-1}(\partial_1 + \mu)w_{11}\rangle \\ & + \langle w_{1+}[(\partial_1)^{-1}w'_{11} - \mu(\partial_1)^{-1}w_{1-} + (\Delta_\mu)^{-1}(\partial_1 - \mu)w'_{11}]\rangle \end{aligned} \tag{12a}$$

It is immediate to notice that there still is one additional multiplier left: w_{1+}. Its associated differential constraint has the solution

$$w_{1-} = (\mu - \partial_1)(\Delta_\mu)^{-1}w'_{11} \tag{11c}$$

Substitution of this value for w_{1-} into the above reduced action leads to the unconstrained light-front action

$$I_{\text{SD,unconstr}} = \langle w'_{11}2\mu^2(\Delta_\mu)^{-2}\dot{w}_{11}\rangle - \langle w_{11}\mu^2(\Delta_\mu)^{-1}w_{11}\rangle \tag{12b}$$

which, after introducing the new variable $w \equiv 2^{(1/2)}\mu\Delta_\mu^{-1}w_{11}$, takes the canonical light-front form

$$I_{\text{SD,unconstr}} = \langle w'\dot{w} - (\tfrac{1}{2})w\Delta_\mu w\rangle \tag{12c}$$

This action naturally leads to the three-dimensional massive Klein-Gordon field equations for the pure (pseudo)spin-2 physical excitation w. Incidentally, observe that the light-front energy, $(\frac{1}{2})\langle w\Delta_\mu w\rangle$, is positive definite.

Now we shift our attention to make the 2 + 1 Hamiltonian analysis of this system. We have to split the different covariant components entering into the action according to their two-dimensional spacelike behavior. In the present case we have that $w_{rs} \equiv (w_{00}, w_{0i}, w_{i0}, w_{ij})$, where $i, j \in (1, 2)$. Two-dimensional objects are decomposed by means of the longitudinal and transverse projectors constructed in the standard way:

$$w_{0i} \equiv \sigma_i v_i^T + \rho_i v_l^L, \qquad w_{i0} \equiv \sigma_i v_2^T + \rho_i v_2^T \tag{13a}$$

$$w_{ij} \equiv \sigma_i\sigma_j w^T + (\rho_i\sigma_j + \rho_j\sigma_i)u^T + 2\rho_i\rho_j u^L + \varepsilon_{ij}h \tag{13b}$$

where $\rho_i \equiv (-\Delta_2)^{-(1/2)}\partial_i$; $\sigma_i \equiv \varepsilon_{ij}\rho_j$; $\rho_i\rho_i = -1 = \sigma_j\sigma_j$; $\rho_i\sigma_i = 0$; $\rho \equiv (-\Delta_2)^{1/2}$.

Computing the action as a function of these $T + L$ components we get that

$$I_{\text{SD}} = (2\mu)^{-1}\langle 2w_{00}[-\rho v_2^T + \mu w^T + 2\mu u^L] + 2v_1^L[\rho u^T - \rho h + \mu v_2^L] + 2v_1^T \rho w^T - 2v_2^L \dot{v}_2^T + 2\mu^T \dot{w}^T + 2h\dot{w}^T + 4u^L\dot{u}^T + 4h\dot{u}^L\rangle - (1/2)\langle 2(u^T)^2 - 2h^2 - 2v_1^T v_2^T - 4w^T u^L\rangle \tag{14a}$$

(the dot stands for time derivatives $\dot{f} \equiv \partial/\partial x^0$).

We observe that w_{00}, v_1^L are Lagrangian multipliers associated with the respective differential constraints

$$u^L = (2\mu)^{-1}\rho v_2^T - (1/2)w^T \tag{15a}$$

$$v_2^L = (\mu)^{-1}(\rho h - \rho u^T) \tag{15b}$$

After plugging these values into the initial constrained action (14a) we obtain a reduced form that contains h as an auxiliary field and yet an additional differential constraint having v_1^T as its multiplier, whose solution is

$$v_2^T = -(\mu)^{-1}\rho w^T \tag{15c}$$

Insertion of this value for v_2^T into the reduced action leads to the final unconstrained form, which can be written in terms of w^T as the unique independent coordinatelike variable and $p \equiv 2\mu^{-1}u^T$ as its canonical associated momenta

$$I_{\text{SD,unconstr}} = \langle pw^T - [(\dot{w}^T)^2 + w^T(-\Delta)(\mu)^{-2}w^T + (\tfrac{1}{4})\mu^2 p^2]\rangle \tag{14b}$$

variations of w^T, p give the Hamiltonian system

$$\dot{p} = -2(\mu)^{-2}[(\mu)^2 - \Delta]w^T \tag{16a}$$

$$\dot{w}^T = (\tfrac{1}{2})\mu^2 p \tag{16b}$$

which again entails the massive Klein–Gordon field equation $(\Box - \mu^2)w^T = 0$ for the unique physical excitation the system has. Note the positive definite character of the energy of the system.

4. VACUUM AMPLITUDE

The last dynamical property we want to consider is the evaluation of the vacuum amplitude of self-dual gravity in the presence of an external source k_{pm}.

In order to do that we have to add to the free action (7c) the external source contribution

$$I_{\text{ext}} \equiv (\mu)^{-1}\langle w_{pm}k_{pm}\rangle \tag{17a}$$

where it is assumed that the external source is conserved, $\partial_p k_{pm} = 0$, in order to keep having the covariant constraint

$$\partial_p w_{mp} = \partial_m w_{pp} \tag{18a}$$

which holds in the free case.

The sourced self-dual massive spin-2 action is defined in the usual way:

$$I_{\mathrm{SD,ext}} \equiv I_{\mathrm{SD}}(7\mathrm{c}) + I_{\mathrm{ext}}(17\mathrm{a}) \tag{17b}$$

which, after independent variations of the graviton field, yields the equations

$$E_{pm} \equiv \varepsilon_p^{\ rs}\partial_r w_{sm} - \mu[w_{mp} - \eta_{mp}w_{rr}] + k_{pm} = 0 \tag{18b}$$

In order to invert this system finding the value of $w_{pm} = w_{pm}(k_{rs})$ it is convenient to introduce covariant algebraic and differential decompositions for the general second-rank tensors as well as for three-dimensional vectors.

Since it is simpler, let us first consider a vector field v_p. Its covariant decomposition is defined to be

$$v_p \equiv v_p^T + \rho_p v^L \tag{19a}$$

$$\rho_p \equiv \partial_p(\square)^{-(1/2)} \tag{19b}$$

$$\rho_p\rho_p = +1$$

where

$$\rho_p v_p^T = 0 \tag{19c}$$

It is immediate to recognize that $v^L = \rho_p v_p$ and $v_p^T = (\eta_{pr} - \rho_p\rho_r)v_r \equiv \pm P_{pr}^T v_r$.

It is interesting to note that, in three dimensions, v_p^T is a reducible projection. In fact v_p^T can be split into its two reflection sensitive parts through the operators $P_r^{\pm s}$

$$v_r^{T\pm} \equiv P_r^{\pm s} v_s^T \equiv (\tfrac{1}{2})[\delta_r^s \pm \varepsilon_r^{ps}\rho_p]v_s^T \tag{19d}$$

$$v_r^T = v_r^{T+} + v_r^{T-} \tag{19e}$$

For a transverse vector v_r^T it is straightforward to check from the above definitions that $v_r^{T\pm}$ are transverse too.

For symmetric second-rank tensors $w_{mp} = w_{pm}$ it is straightforward to make the $T + L$ decomposition. It turns out to be

$$w_{pm} \equiv w_{pm}^{Tt} + (\eta_{pm} - \rho_p\rho_m)w^T + \rho_p w_{2m}^T + \rho_m w_{2p}^T + \rho_p\rho_m w^L \tag{20a}$$

where $\rho_p w_{pm}^{Tt} = 0 = w_{pp}^{Tt}$, $\rho_r w_{2r}^T = 0$. The transverse-traceless part contains two degrees of freedom, w_m^T has another two, and w^T and w^L complete the six independent variables needed to represent a three-dimensional symmetric second-rank tensor.

As it happens for the vector case, w_{pm}^{Tt} can still be split into its two reflection-sensitive parts $w_{pm}^{Tt\pm}$ by introducing the pseudo-spin ±2 projectors on the Tt subspace,

$$\begin{aligned} w_{pm}^{Tt\pm} &\equiv P_{pm}^{rn\pm} w_{rn}^{Tt} \\ &= (\tfrac{1}{2}) w_{pm}^{Tt} \pm (\tfrac{1}{4})(\varepsilon_p^{\ sr} \rho_s w_{rm}^{Tt} + \varepsilon_m^{\ sr} \rho_s w_{rp}^{Tt}) \end{aligned} \tag{20b}$$

These two projectors satisfy that

$$\begin{gathered} P_{pm}^{rn+} + P_{pm}^{rn-} = 1_{pm}^{rnTt}, \qquad (P_{pm}^{rn\pm})^2 = P_{pm}^{rn\pm} \\ \rho_p w_{pm}^{Tt\pm} = 0 \end{gathered} \tag{20c}$$

If one decomposes both the source and the spin-2 field into their respective symmetric and antisymmetric components

$$k_{pm} \equiv \bar{k}_{pm} + \varepsilon_{pmt} k_{1t}, \qquad w_{pm} \equiv \bar{w}_{pm} + \varepsilon_{pmt} w_{1t} \tag{20d}$$

one can write the conservation law in the form

$$k_2^L = 0, \qquad k_{2m}^T = \varepsilon_m^{\ rs} \rho_r k_{1s}^T \tag{21a}$$

and the covariant constraint (18a) in a new way

$$w^T = 0, \qquad w_{2m}^T = -\varepsilon_m^{\ rs} \rho_r w_{1s}^T \tag{21b}$$

where k_{1s}^T, w_{1s}^T are the transverse components of k_{1s}, w_{1s}.

From the antisymmetric projections of equation (18b) we obtain that w_{ir} does not propagate:

$$w_{1r} = -(\mu)^{-1} k_{1r} \tag{22a}$$

The trace of the same system gives for w_2^L

$$w_2^L = -(\mu)^{-1} k^T - \rho(\mu)^{-2} k_1^L \tag{22b}$$

Taking the remaining projections of the symmetric part of (18) we have for the remaining intrinsic components of w_{pm}

$$w_{2p}^{T\pm} = \pm(\mu)^{-1} k_{1p}^{T\pm}, \qquad w_{pm}^{Tt\pm} = \mp(\rho \pm \mu)(\Box - \mu^2)^{-1} k_{pm}^{Tt\pm} \tag{22c}$$

In a complexive way all this information says that the symmetric part w_{pm} of the graviton field has, in terms of the source, the value

$$\begin{aligned} \bar{w}_{mp} = -(\Box - \mu^2)^{-1}[&\mu \bar{k}_{mp} - (\tfrac{1}{2})\mu \eta_{mp} \bar{k}_{rr} + (\tfrac{1}{2})\varepsilon_m^{\ rs} \partial_r \bar{k}_{sp} \\ &+ (\tfrac{1}{2})\varepsilon_p^{\ rs} \partial_r \bar{k}_{sm} + (2\mu)^{-1} \partial_{mp} \bar{k}_{rr} + (\mu)^{-2} \partial_{mpr} kn_{1r} \\ &- (\tfrac{1}{2})\partial_m k_{1p} - (\tfrac{1}{2})\partial_p k_{1m} \\ &- (\mu)^{-1} \varepsilon_m^{\ rs} \partial_{pr} k_{1s} - (\mu)^{-1} \varepsilon_p^{\ rs} \partial_{mr} k_{1s}] \end{aligned} \tag{22d}$$

Making use of the covariant constraint equation (18a) $I_{\text{SD ext}}$ can be seen to be

$$I_{\text{SD ext}}|_{(18a)} = \langle \bar{w}_{mp}\varepsilon^{prs}\partial_r \bar{w}_{ms}\rangle + (2\mu)^{-1}\langle w_{1p}\varepsilon^{prs}\partial_r w_{1s}\rangle - (\tfrac{1}{2})\langle \bar{w}^2_{mp} - \bar{w}^2_{rr}\rangle - \langle w^2_r\rangle + (\mu)^{-1}\langle w_{mp}\bar{k}_{mp} - 2k_{1r}w_{1r}\rangle \tag{23a}$$

We introduce in this action the values (22a) and (22d) already found for the spin-2 field in terms of the source. We obtain that

$$\begin{aligned} I_{\text{SDext}}|_{(18)} = \langle &-(\tfrac{1}{2})\bar{k}_{mp}(\Box - \mu^2)^{-1}\bar{k}_{mp} - \bar{k}_{mp}(\Box - \mu^2)^{-1}\varepsilon_p^{rs}(2\mu)^{-1}\partial_r \bar{k}_{ms} \\ &+ (\tfrac{1}{4})\bar{k}_{pp}(\Box - \mu^2)^{-1}\bar{k}_{rr} - k_{1r}(\Box - \mu^2)^{-1}k_{1r} \\ &- k_{1p}(\Box - \mu^2)^{-1}\varepsilon_p^{rs}(2\mu)^{-1}\partial_r k_{1s} - \mu^{-2}\partial_r k_{1r}(\Box - \mu^2)^{-1}\partial_s k_{1s}\rangle \end{aligned} \tag{23b}$$

In terms of the intrinsic components of the source this action can be written in a physically more appealing form:

$$I_{\text{SDext}}|_{(18)} = -(\tfrac{1}{2})\langle k^{Tt}_{mp}(\Box - \mu^2)^{-1}[k^{Tr}_{mp} + \varepsilon^{rs}_m(\mu)^{-1}\partial_r k^{Tt}_{sp}\rangle - (\mu)^{-2}\langle k^L_1 k^L_1\rangle \tag{23c}$$

If, in order to sharpen our physical interpretation, we allow the exchange of μ for p in the reflection sensitive part of the Tt-quantum exchange, we recognize that (23c) becomes

$$I_{\text{SDext}} = -\langle k^{Tt+}_{mp}(\Box - \mu^2)^{-1}k^{Tt+}_{mp}\rangle - (\mu)^{-2}\langle k^L_1 k^L_1\rangle \tag{23d}$$

showing in a transparent way that the vacuum amplitude is given by the exchange of quanta of a definite pseudospin (plus an irrelevant contact term).

5. CONCLUSIONS

We have shown that the third-order topologically massive three-dimensional graviton, if linearized, can be equivalently described by a first-order self-dual-like action.

This self-dual spin-2 action does not have any of the two gauge invariances that the topologically massive action possesses. First-order master actions that lead to the intermediate second-order system and to the topological third-order one were given. All of them, when some of the generated field equations are used, collapse into the self-dual massive spin-2 action.

Then we show both in the light-front representation and in the more conservative time + space Hamiltonian formulation that the respective unconstrained actions depend upon only one dynamical variable (not the same), which are properly located so as to lead us naturally believe they are the right pseudospin 2 variables [5].

Both the light-front and the timelike Hamiltonians are explicitly exhibited showing their positive definite structure.

The basic quantum elements, the propagator and the vacuum amplitude in the presence of an external conserved source, are computed in a covariant form. The vacuum amplitude can be transparently interpreted as corresponding (up to an irrelevant contact term) to the exchange of the covariant pseudospin ± 2 excitations, the unique physical degree of freedom kept alive in three-dimensional gravity.

Some questions concerning this model are worth considering, viz., what the supersymmetric version of this bosonic action looks like and whether this intrinsically first-order action can be consistently coupled to gauge invariant topologically massive gravity.

ACKNOWLEDGMENT. This research was partially supported by ROSTLAC grant No. 211-848-5 and NSF grant No. PHY 82-01094.

REFERENCES

1. S. Deser and R. Jackiw, *Phys. Lett. 139B*, 371 (1984).
2. P. K. Towsend, K. Pilch, and P. van Nieuwenhuizen, *Phys. Lett. 136B*, 38 (1984).
3. W. Siegel, *Nucl. Phys. B156*, 135 (1979); R. Jackiw and S. Templeton, *Phys. Rev. D 23*, 2291 (1981); J. Schonfeld, *Nucl. Phys. B185*, 157 (1985).
4. C. Aragone and A. Khoudeir, Self-dual massive gravity, SB/FM/F-125. *Phys. Lett. 173B*, 141 (1986).
5. S. Deser, R. Jackiw, and S. Templeton, *Phys. Rev. Lett. 48*, 975 (1982); *Ann. Phys. (N.Y.) 140*, 372 (1982).

Chapter 3

Non-Abelian Chern–Simons Topological Coupling from Self-Interaction

Carlos Aragone and J. Stephany

At present, unified theories of all the fundamental interactions (including gravity) based on superstring theories are receiving much attention.

The more interesting of them are anomaly free if the gauge group is $SO(32)$ or $E_8 \times E_8$ [2] and have (all three of them) as their low-energy limit $N = 1$, $d = 10$ Einstein-Yang-Mills supergravity (EYMS) [3].

The $N = 1$, $d = 10$ EYMS fundamental supergravity multiplet is $\{e_{ma}, \Psi_m, b_{mn}, \lambda, \phi\}$: the zehnbein, gravitino, second-rank antisymmetric tensor, Dirac field, and a scalar, respectively. The Yang-Mills supermultiplet is $\{A^a_m, \chi^a_{\cdot}\}$: the non-Abelian vector and the Dirac field in the adjoint representation of the gauge group, respectively. The unification of Yang–Mills theory and supergravity in $d = 10$ [3] was obtained by an adequate generalization of the $N = 1$, $d = 10$ Einstein–Maxwell supergravity [4].

The central results of these constructions are how the second-rank antisymmetric tensor turns out to be coupled to the gauge vector field, both for the Maxwell and for the Yang–Mills systems.

Here we present recent results showing [1] that the (second-rank) antisymmetric tensor-Chern–Simons type coupling to the non-Abelian vector field can be obtained starting from the antisymmetric Maxwell

CARLOS ARAGONE and J. STEPHANY • Department of Physics, Simón Bolívar University, Caracas 1080-A, Venezuela.

coupling just as the immediate consequence of the self-interaction mechanism [5]. The result is very much similar to what was already found for the three-dimensional topological massive Yang–Mills system [6]: non-Abelian Chern–Simons terms grow up from the differential topological-like terms appearing in the associated Abelian model through the self-interaction mechanism, the difference between the three-dimensional case and the higher-dimensional ones ($d = 10$, for instance) stemming from the fact that in the former case the Chern–Simons term appears in an integral form (it constitutes the mass-generating contribution to the action), while in the latter its presence occurs in a local way, appearing as a kind of topological interaction that couples quadratically and in a cubic way the antisymmetric tensor to the Yang–Mills vector field.

We start from the first-order formulation of the second-rank antisymmetric tensor coupled to a Maxwell field

$$I_d(A, a, f, G) \equiv -\tfrac{3}{2}\langle[a_{mnp} - \kappa(A_m A_{np} + A_n A_{pm} + A_p A_{mn})]G^{mnp}\rangle + \tfrac{3}{4}\langle G^2_{mnp}\rangle - \tfrac{1}{2}\langle A_{np} f_{np}\rangle + \tfrac{1}{4}\langle f^2_{np}\rangle \tag{1}$$

where the angle brackets indicate the d-dimensional integration, κ is the d-dimensional gravitational coupling constant, and $A_{np} \equiv \partial_n A_p - \partial_p A_n$, $a_{mnp} \equiv \partial_m a_{np} + \partial_n a_{pm} + \partial_p a_{mn}$ are the two field strengths, invariant under their respective Abelian gauge transformations:

$$\delta_\chi A_r = \partial_r \chi \tag{2a}$$

$$\delta_\xi a_{np} = \partial_n \xi_p - \partial_p \xi_n \tag{2b}$$

The independent variables (f_{np}, G_{mnp}) are also assumed to be gauge invariant:

$$\delta_\xi G_{mnp} = \delta_\chi G_{mnp} = 0 \tag{2c}$$

$$\delta_\xi f_{np} = 0 = \delta_\chi f_{np} \tag{2d}$$

The quadratic coupling—$\kappa A\, \partial A$ of the antisymmetric tensor—does not break the gauge invariance of the full action (1) it is also transforms under (2a) according to

$$\delta_\chi a_{np} = \kappa \chi A_{np} \tag{2e}$$

as was shown by Bergshoeff *et al.* [4].

Independent variations of (f_{np}, G_{mnp}) yield their respective values

$$f_{np} = A_{np}, \qquad G_{mnp} = a_{mnp} - 3\kappa A_{[m} A_{np]} \tag{3}$$

Which, if introduced into the action (1), lead to the original second-order form of Bergshoeff *et al.* [4].

Now consider (for simplicity) a triplet of Abelian vector fields A_r^a, $a = (1, 2, 3)$ coupled with the antisymmetric tensor a_{np}. The action of this system is the addition of the three Maxwellian contributions to the action of the free antisymmetric tensor:

$$I_d^{\text{Abelian}}(A_r^a, a, f_{rs}^a, G) \equiv -\tfrac{3}{2}\langle [a_{mnp} - 3\kappa A_{[m}^a A_{np]}^a] G_{mnp}\rangle + \tfrac{3}{4}\langle G_{mnp}^2\rangle - \tfrac{1}{2}\langle A_{np}^a f_{np}^a\rangle + \tfrac{1}{4}\langle (f_{np}^a)^2\rangle \tag{4}$$

This action is gauge invariant under the natural generalization of (2a) and (2e)

$$\delta_\chi A_r^a = \partial_r \chi^a \tag{2a'}$$

$$\delta_\chi a_{np} = \kappa \chi^a A_{np}^a \tag{2e'}$$

and presents a rigid $SU(2)$ invariance:

$$\delta_\omega A_r^a = [A_r \times \omega]^a \equiv \varepsilon_{bc}^a A_r^b \omega^c \tag{5a}$$

$$\delta \omega f_{nr}^a = [f_{nr} \times \omega]^a \tag{5b}$$

[a_{np}, G_{mnp} are $SU(2)$ scalars].

This rigid invariance ensures the existence of an associated conserved Noether current

$$j_a^m \equiv \frac{\delta I}{\delta(\partial_m A_n^b)} \frac{\delta A_n^b}{\delta \omega^a} = [f_{mn} \times A_n]_a - 9\kappa G_{mnp}[A_n \times A_p]_a \tag{5c}$$

The self-interaction mechanism consists in making j_a^m the intrinsic course of the Maxwell equations:

$$\frac{\delta I^{\text{Abelian}}}{\delta A_m^a} = j_a^m \tag{6a}$$

In order to obtain this field equation, one has to augment the initial Abelian action into a new term, $I^{\text{non-Abelian}}$ such that

$$\frac{\delta I^{\text{non-Abelian}}}{\delta A_m^a} = -j_a^m[(5c)] \tag{6b}$$

It is very simple to integrate this functional differential equation,

$$I^{\text{non-Abelian}} = -\tfrac{1}{2}\langle \varepsilon^{abc} A_m^a A_n^b f_{mn}^c\rangle + 3\kappa\langle G_{mnp}\varepsilon^{abc} A_m^a A_n^b A_p^c\rangle \tag{6c}$$

the new action $I_d^{\text{YM}}(A_r^a,\, a,\, f_{rs}^a G)$ defined by

$$I_d^{\text{YM}} \equiv I^{\text{Abelian}} + I^{\text{non-Abelian}} \tag{7a}$$

is easily seen to have the value

$$I_d^{\mathrm{YM}}(A_r^a, a_{np}, f_{rs}^a, G_{mnp})$$
$$= -\tfrac{3}{2}\langle[a_{mnp} - 3\kappa A^a_{[m} f^a_{np]}]G^{mnp}\rangle$$
$$+ \tfrac{3}{4}\langle G^2_{mnp}\rangle - \tfrac{1}{2}\langle(A_{np} + A_n \times A_p)^a f^a_{mn}\rangle + \tfrac{1}{4}\langle(f^a_{mn})^2\rangle \tag{7b}$$

where we have introduced the distorted Yang–Mills field strength (which is not a local isovector)

$$f^a_{np}(A) \equiv A^a_{np} + \tfrac{2}{3}(A_n \times A_p)^a = f^a_{np}(A) - \tfrac{1}{3}(A_n \times A_p)^a \tag{8}$$

This is the Chapline and Manton [3] action in its first-order form, which has been obtained in only one step after postulating the self-interaction of the initially Abelian vector fields, already topologically coupled to the antisymmetric gauge matter

Since the self interaction also leads to the fusion of the two separate sets of parameters $\chi(x)$, ω^a into a unique set $\omega^a(x)$, we have to check the consistency of the new action (7b) under the local gauge transformations that result from this fusion. They are

$$\delta_\omega A^a_r = D_r\omega^a \equiv [\partial_r \omega]^a + [A_r \times \omega]^a \tag{9a}$$

$$\delta_\omega f^a_{rs} = [f_{rs} \times \omega]^a \tag{9b}$$

$$\delta_\omega a_{np} = \kappa\omega^a A^a_{np} + \lambda_{np}(\omega; A) \tag{9c}$$

where λ has to be determined. After realizing that

$$\delta_\omega\{a_{mnp} - 3\kappa A^a_{[m} f^a_{np]}\} = 3\kappa\lambda\,\partial_{[m}\lambda_{np]} \tag{9d}$$

one finally obtains that $\lambda_{np} = \partial_{[n}\lambda_{p]}$ if G_{mnp} must be gauge invariant; this part of the transformation can always be absorbed by an appropriate Abelian transformation (2b).

It is worth mentioning that, under a large (finite) gauge transformation, it has recently been found [7] that a_{np} picks out, in addition to the finite form of (9c), an additional finite contribution that has the role of matching the finite transformation law of the Chern–Simons part of G_{mnp}.

There is an alternative form of obtaining the non-Abelian Chern-Simons topological coupling, starting from the dual formulation of the Abelian action [5], as was given by Baulieu [8] following the dualization methods proposed earlier by Cremmer, Julia, Nicolai, and Townsend [9]. In its second-order form it is

$$I_{\mathrm{dual}}(A_r, h_{mnp}, m_{r_1\ldots r_6}) \equiv -\tfrac{1}{4}\langle A^a_{mn}(A)^2\rangle - \tfrac{3}{4}\langle h^2_{mnp}\rangle + \frac{1}{6(6!)}$$
$$\times\langle\varepsilon^{r_1 r_2\cdots r_9 r_{10}} m_{r_1\ldots r_6}[\partial_{[r_7} h_{r_8 r_9 r_{10}]} - \tfrac{3}{2}A^e_{[r_7 r_8} A^a_{r_9 r_{10}]}] \tag{10a}$$

This action is gauge invariant under the Abelian group of transformations $\delta A_r^a = \partial_r \chi^a \delta m_{r_1 \ldots r_6} = \partial_{r_1} \lambda_{r_2 r_3 \ldots r_{10}} - \partial_{r_2} \lambda_{r_3 \ldots r_6 r_1} + \cdots - \partial_{r_6} \lambda_{r_1 \ldots r_5}$.

The local equivalence of this action with (4) arises through independent variations of $m_{r_1 r_2 \ldots r_6}$ which lead to the field equation:

$$dh_3 = A_2^2 \tag{11}$$

where we are using the geometric notation and the exterior calculus for simplicity [10]: $h_3 \equiv (3!)^{-1}\, h_{mnp} dx^{mnp}$; $A_2 \equiv (2!)^{-1}\, A_{rs} dx^{rs}$. Since $dA_2^2 = 0$, equation (11) is locally equivalent to (3) and the insertion of this solution into the above action (10a) gives the second-order version of (4)

$$I_{\text{dual}}(A_r, G_{mnp}(A_r, a_{mn})) = -\tfrac{1}{4}\langle (A_{mn}^a)^2 \rangle - \tfrac{3}{4}\langle G_{mnp}^2(A_r, a_{mn})\rangle \tag{10b}$$

The first-order formulation of action (10a) can be obtained by appropriate cloning of the terms quadratics in the vector field strengths A_{mn}^a. It turns out to be

$$\begin{aligned} I_{\text{dual}}&(A_r, f_{mn}, h_{mnp}, m_{r_1 \ldots r_6}) \\ &= \langle -\tfrac{1}{2} A_{mn}^a f_{mn}^a + \tfrac{1}{4}(f_{mn}^a)^2 \rangle \\ &\quad - \tfrac{3}{4}\langle h_{mnp}^2 \rangle + \frac{1}{6(6!)} \langle \varepsilon^{r_1 \cdots r_6 r_7 \cdots r_{10}} m_{r_1 \ldots r_6} [\partial_{[r_7} h_{r_8 r_9 r_{10}]} - 3A_{[r_7 r_8}^a f_{r_9 r_{10}]}^a \\ &\quad + \tfrac{3}{2} f_{[r_7 r_8} f_{r_9 r_{10}]}] \rangle \end{aligned} \tag{12}$$

Independent variations of f_{mn}^a yield

$$\left(\frac{1}{2} \delta_{mn}^{pq} + \frac{1}{6!} \varepsilon_{mn}^{pq r_5 \cdots r_{10}} m_{r_5 \ldots r_{10}} \right) (f_{pq}^a - A_{pq}^a) = 0 \tag{13}$$

which lead to $f_{pq}^a = A_{pq}^a$ as required in order to recover (10a) when this value of f_{pq}^a is introduced into (12).

Now it is immediate to saturate this action through the self-interaction mechanism, since this action (12) is linear in the first-order derivatives of the gauge vector fields.

The Noether current has the value

$$j_{\text{dual}m}^a = -[f_{mn} \times A_n]^a - \tfrac{1}{6!} \varepsilon^{r_1 \cdots r_6 m n r_9 r_{10}} m_{r_1 \ldots r_6} [f_{r_9 r_{10}} \times A_n]^a \tag{14}$$

which, being linear in A_n^b, after functional integration with respect to A_r^b gives the non-Abelian part of the initial Abelian system:

$$\begin{aligned} I^{\text{non-Abelian}} &\equiv -\tfrac{1}{2}\langle \varepsilon^{abc} A_m^a A_n^b f_{mn}^c \rangle \\ &\quad - \frac{1}{2(6!)} \langle \varepsilon^{r_1 \cdots r_6 m n r_9 r_{10}} m_{r_1 \ldots r_6} \times \varepsilon^{abc} A_m^a A_n^b f_{mn}^c \rangle \end{aligned} \tag{15}$$

Therefore, the final saturated first-order action of this dualized Chapline-Manton system is the addition of the Abelian leading part plus the non-Abelian self-coupled part (15) stemming in the self-interaction mechanism:

$$
\begin{aligned}
I^{\mathrm{YM}}_{d,\mathrm{dual}}&(A^a_r, a_{np}, f^a_{rs}, h_{mnp}, m_{r_1 \ldots r_6}) \\
&= \langle -\tfrac{1}{2}[\partial_m A^a_n - \partial_n A^a_m + \varepsilon^{abc} A^b_m A^c_n] f^a_{mn} + \tfrac{1}{2}(f^a_{mn})^2\rangle - \tfrac{3}{4}\langle h^2_{mnp}\rangle \\
&\quad + \tfrac{1}{6(6!)}\langle \varepsilon^{r_1 \cdots r_6 \cdots r_{10}} m_{r_1 \ldots r_6}[\partial_{[r_7} h_{r_8 r_9 r_{10}]} \quad - 3[\partial_{[r_7} A^a_{r_8} - \partial_{r_8} A^a_{f_7} \\
&\quad + \varepsilon^{abc} A^b_{r_7} A^c_{r_8}] f^a_{r_9 r_{10}]} + \tfrac{3}{2} f^a_{r_7 r_8} f^a_{r_9 r_{10}]}]\rangle
\end{aligned} \tag{16}
$$

This action is locally equivalent to our previous action (7b).

ACKNOWLEDGMENT. This research was partially supported by CONICIT grant No. S1-972 and ROSTLAC grant No. 211-48-5.

REFERENCES

1. C. Aragone and J. E. Stephany, *Phys. Rev. D 34*(4), 1210 (1986).
2. M. B. Green and J. H. Schwarz, *Phys. Lett. 149B* 117 (1984); D. J. Gross, J. Harvey, E. Martinec, and R. Rohm, *Phys. Rev. Lett. 54*, 502 (1985).
3. G. F. Chapline and N. S. Manton, *Phys. Lett. 120B*, 105 (1983).
4. E. Bergshoeff, M. de Roo, B. de Wit, and P. van Nieuwenhuizen, *Nucl. Phys. B195*, 97 (1982); A. H. Chamseddine, *Nucl. Phys. B185*, 403 (1981).
5. S. Deser, *Gen. Rel. Grav. 1*, 9 (1970).
6. C. Aragone and G. Araujo, *Acta Cient. Ven. 36*, 207 (1985).
7. R. I. Nepomechie, Y. S. Wu, and A. Zee, *Phys. Lett. 158B*, 311 (1985).
8. L. Baulieu, *Nucl. Phys. B227*, 157 (1983).
9. H. Nicolai and P. Townsend, *Phys. Lett. B98*, 257 (1981); G. Cremmer and B. Julia, *Nucl. Phys. B159*, 141 (1979).
10. B. Zumino, Y. S. Wu, and A. Zee, *Nucl. Phys. B229*, 449 (1984).

Chapter 4

Grassmannian Space-Time Structure as an Origin of Gauge Symmetries

Laurent Baulieu

The simplifications brought into the analysis of gauge symmetries acting on local fields by enlarging the Minkowski space-time $\{x^\mu\}$ into a space with local coordinates $\{x^\mu, \theta, \bar{\theta}\}$, where θ and $\bar{\theta}$ are a pair of Grassmannian coordinates with no spinorial charge, constitute a very intriguing phenomenon. In this framework, the BRST and anti-BRST operators are naturally introduced as exterior differential operators along the unphysical coordinates θ and $\bar{\theta}$, and the requirement of BRST invariance, i.e., of gauge independence of physics, means simply that physics must be invariant under displacements along the unphysical directions [1]. It must be understood that the space $\{x^\mu, \theta, \bar{\theta}\}$ is the same for all types of gauge symmetries, and constitutes therefore a pure kinematical entity, independent of the notion of a gauge field. The space $\{x^\mu, \theta, \bar{\theta}\}$ has effective dimension $(D-2)$, since the θ and $\bar{\theta}$ coordinates should be counted negatively, as violating the physical statistics [1, 2]. This is a further indication, in our belief, that it constitutes an appropriate kinematical framework for the description of massless gauge fields. The use of this space has already provided simple descriptions of these gauge symmetries which involve p-form gauge

LAURENT BAULIEU • Laboratoire de Physique Théoretique et Hautes Energies, Université Pierre and Marie Curie, 75252 Paris Cedex 05, France.

fields ($p \geq 1$),

$$B_p = \frac{1}{p!} B_{[\mu_1 \ldots \mu_p]} \, dx^{\mu_1} \wedge \cdots \wedge dx^{\mu_p}$$

and generalize straightforwardly the Yang–Mills symmetry [1]. The symmetry under local diffeormorphisms can be included in a natural way in the formalism, and a simple algebraic description of gravity and supergravities has also been obtained [1, 2].

In this chapter I will briefly recall the known results concerning the determination of gauge symmetries in flat space from generalized Maurer-Cartan equations. Then I will point out a new result, namely, that these equations can be obtained from an action principle in the enlarged space $\{x^\mu, \theta, \bar{\theta}\}$.

Gauge fields are introduced as a set of exterior forms $\tilde{B}_{p_i}(x, \theta, \bar{\theta})$ ($p_i \geq 1$) classified by the index i, and possibly valued in given representations of Lie algebras. The value $p = 1$ corresponds to the Yang–Mills case. The space p of differential forms is spanned by the basis of forms $\alpha^{q,\bar{q}}_{K-q-\bar{q}}$

$$\alpha^{q,\bar{q}}_{K-q-\bar{q}} = dx^{\mu_1} \wedge \cdots \wedge dx^{\mu_{K-q-\bar{q}}} \wedge (d\theta)^q \wedge (d\bar{\theta})^{\bar{q}}$$

$$(d\theta)^q \equiv \underbrace{d\theta \wedge \cdots \wedge d\theta}_{q}$$

$$(d\bar{\theta})^{\bar{q}} \equiv \underbrace{d\bar{\theta} \wedge \cdots \wedge d\bar{\theta}}_{\bar{q}}$$

Notice that p is infinite dimensional because $(d\theta)^q \wedge (d\bar{\theta})^q$ never vanishes whatever the values of q and $\bar{q}$ are. Indeed, $d\theta$ and $d\bar{\theta}$ are exterior forms of Grassmannians, and thus are commuting objects. One can expand $\tilde{B}_p$ as

$$\tilde{B}_p(x, \theta, \bar{\theta}) = B_p(x, \theta, \bar{\theta}) + \sum_{g=1}^{p} \sum_{\substack{q+\bar{q}=g \\ q,\bar{q}\geq 0}} B^{q,\bar{q}}_{p-g}(x, \theta, \bar{\theta}) \tag{1}$$

with

$$B^{q,\bar{q}}_{p-g}(x, \theta, \bar{\theta}) = \frac{1}{(p-g)!} \frac{1}{q!} \frac{1}{\bar{q}!} B^{q,\bar{q}}_{\mu_1 \ldots \mu_{p-g}}(x, \theta, \bar{\theta}) \, dx^{\mu_1} \wedge \cdots \wedge dx^{\mu_{p-g}}$$

B_p, which has no $d\theta$ or $d\bar{\theta}$ component, is called by definition the classical p-form gauge field, and $B^{q,\bar{q}}_{p-g}$, $g = q + \bar{q} \geq 1$ is the ghost of order g of B_p, with antighost number $\bar{q}$. The statistics of any object is defined as the sum of its Lorentz degree and ghost number. One postulates that a p-form gauge field is described in ordinary field theory by all components $B^{q,\bar{q}}_{\mu_1 \ldots \mu_g}(x, \theta, \bar{\theta}) \times (d\theta)^q \wedge (d\bar{\theta})^{\bar{q}}$, classical and ghost, contained in the generalized form $\tilde{B}_p$, and evaluated at $\theta = \bar{\theta} = 0$. This ansatz is motivated from the necessity of introducing a system of fields carrying effectively $\binom{P}{D-2}$ degrees of freedom which is the number of physical degrees of freedom described by a classical p-form massless gauge field in D-dimensional Minkowski space-time [1].

To determine the variations of all fields contained in $\tilde{B}_{p-g}(x, \theta, \bar{\theta})$ one introduces the exterior differential operator $\tilde{d}$ in P

$$\begin{aligned} \tilde{d} &= d + s + \bar{s} \\ d &\equiv dx^{\mu} \frac{\partial}{\partial x^{\mu}} \\ s &\equiv d\theta \frac{\partial}{\partial \theta}, \qquad sd + ds = 0 \\ \bar{s} &= d\bar{\theta} \frac{\partial}{\partial \bar{\theta}}, \qquad \bar{s}d + d\bar{s} = 0 \end{aligned} \tag{2}$$

At $\theta = \bar{\theta} = 0$ one identifies the action of s and $\bar{s}$ on all fields as the action of the BRST and anti-BRST operators, respectively. This definition allows one to interpret infinitesimal gauge transformation as displacements along the unphysical directions. The basic rule of exterior calculus is $\tilde{d}^2 = 0$. By expansion in ghost number this yields $sd + ds = \bar{s}d + d\bar{s} = 0$ and $s^2 = s\bar{s} + \bar{s}s = \bar{s}^2 = 0$. The latter property corresponds in fact to the closure and Jacobi relation of the system of gauge transformations associated with s and $\bar{s}$ [3].

Using the operator $\tilde{d}$, we can build the field strength $\tilde{G}^i_{p_i+1}$ of the "potential" $\tilde{B}_{p_i}$.

$$\tilde{G}^i_{p_i+1}(x, \theta, \bar{\theta}) = \tilde{d}\tilde{B}^i_{p_i} + R^i_{p_i+1}(\tilde{B}, \tilde{G}) \tag{3}$$

$\tilde{d}\tilde{B}^i_{p_i}$ is the free (i.e., Maxwell) part of the field strength. The interacting part R^i_{p+1} is an exterior $(p_i + 1)$-form function of the potentials $\tilde{B}^K$ and of their exterior derivatives, that is to say a function of $\tilde{B}^{(K)}$ and $\tilde{G}^{(K)}$ with $K \neq i$. R_{p+1} must be such that the $\tilde{G}_{p+1}$'s satisfy Bianchi identities [1, 2]. This means that one must have

$$\tilde{d}\tilde{G}^i_{p_i+1} = \sum_j d^{ij}_{p_i-p_j}(\tilde{G}^K, \tilde{B}^K)\tilde{G}^j_{p_j+1} \tag{4}$$

The $(p_i - p_j)$-form d^{ij} is an exterior product of $\tilde{G}_K$ and $\tilde{B}_K$ such that its dependence on the potentials $\tilde{B}_K$, at fixed $\tilde{G}_{K+1}$ is only through the 1-form (i.e., Yang-Mills) gauge fields contained in the set of fields $\tilde{B}$, and not more than linearly [1].

The field strengths $\tilde{G}_{p+1}$ contains the θ and $\bar{\theta}$ variations of all classical and ghost components of $\tilde{B}_p$, through the term $(s + \bar{s})B_p$. To determine these variations one imposes that all terms in $\tilde{G}_{p+1}$ with nonzero ghost number vanish. In other words, we impose that $\tilde{G}_{p+1}$ satisfies a

Maurer-Cartan-type equation [1, 2]

$$\begin{gathered}\tilde{G}_{p+1}(x, \theta, \bar{\theta}) = \tilde{G}_{\mu_1\cdots\mu_{p+1}}\, dx^{\mu_1} \wedge \cdots \wedge dx^{\mu_{p+1}} \equiv G_{p+1} \\ \Leftrightarrow \tilde{G}^{q,\bar{q}}_{p-g} = 0, \qquad g = q + \bar{q} \geq 1\end{gathered} \tag{5}$$

In fact, expanding equation (5) in ghost number, one gets for the maximal (i.e., $q = 0$ or $\bar{q} = 0$) ghosts and classical fields (i.e., $g = 0$)

$$\begin{aligned} sB^{g,0}_{p-g} &= -dB^{q+1,0}_{p-g-1} - [R_{p+1}(\tilde{B}, G)]^{g+1,0}_{p-g} \\ \bar{s}B^{0,g}_{p-g} &= -dB^{0,g+1}_{p-g-1} - [R_{p+1}(\tilde{B}, G)]^{0,g+1}_{p-g} \end{aligned} \tag{6}$$

For the inner ghosts ($q \neq 0$ or $\bar{q} \neq 0$), one has

$$sB^{q,\bar{q}}_{p-g} + \bar{s}B^{q+1,q-1}_{p-g} = -dB^{q+1,\bar{q}}_{p-g-1} - [R_{p+1}(\tilde{B}, G)]^{q+1,\bar{q}}_{p-g} \tag{7}$$

This degeneracy of the action of s and $\bar{s}$ upon inner ghosts is not essential. It is raised by introducing a set of auxiliary fields $b^{q,\bar{q}}_{p-1-g}$

$$\tilde{b}_p = b_p + \sum_{g=1}^{p-1} \sum_{g=q+\bar{q}\geq} b^{q,\bar{q}}_{p-1-g} \tag{8}$$

and by defining ($g \geq \bar{q} \geq 1$)

$$\begin{aligned} sB^{q,\bar{q}}_{p-g} &= b^{q,\bar{q}-1}_{p-g} \\ sb^{q,\bar{q}-1}_{b-g} &= 0 \\ \bar{s}B^{q+1,\bar{q}-1}_{p-g} &= -b^{q,\bar{q}-1}_{p-g} - dB^{q+1,\bar{q}}_{p-g-1} - [R_{p+1}(\tilde{B}, G)]^{q+1,\bar{q}}_{p-g} \\ \bar{s}b^{q,\bar{q}-1}_{p-g} &= d(\bar{s}B^{q+1,\bar{q}}_{p-g-1}) - \bar{s}[R_{p+1}(\tilde{B}, G)]^{q+1,\bar{q}}_{p-g} \end{aligned} \tag{9}$$

In Hamiltonian language, one can interpret the components of $\tilde{b}$ as momenta of temporal components of classical and ghost fields [4].

The Bianchi identity (4) guarantees that s and $\bar{s}$, determined from equations (6) and (9), are nilpotent:

$$s^2 = s\bar{s} + \bar{s}s = \bar{s}^2 = 0 \tag{10}$$

Explicit examples of this construction can be found in Refs. 2 and 5. Invariant classical Lagrangians are of the type

$$\mathcal{L} = \sum_i G^i_{p_i+1} * G^i_{p_i+1} \tag{11}$$

Apart from technicalities, the interesting points of the above determination of gauge symmetries construction are as follows. One has considered as a kinematical space the space with local coordinates $\{x^\mu, \theta, \bar{\theta}\}$, which has an effective dimension equal to $D - 2$, owing to the Grassmann character of the coordinates θ, $\bar{\theta}$, called unphysical by definition. In this space, a p form has unphysical, i.e., ghost, components along the θ and $\bar{\theta}$ directions. This is in fact admissible because a potential is not observable. On the other hand, the field strength of a potential, defined as linear in the derivative of the potential, plus some interactions compatible with Bianchi identities, must be observable, and therefore without unphysical components. This leads to the Maurer–Cartan-type equations (5), i.e., to the BRST equations (6) and (9), which express in fact the action of gauge transformations on all the fields, classical and ghost, contained in $\tilde{B}_p$. One can also interpret the BRST equations under the form (5) as differential equations in θ and $\bar{\theta}$, which allows one to integrate out the θ and $\bar{\theta}$ dependence, and make them unobservable. In any case, the use of the space $\{x, \theta, \bar{\theta}\}$ for describing local gauge fields in the framework of a differential structure, while enforcing the property that the θ and $\bar{\theta}$ coordinates are unphysical, leads us without ambiguity to the horizontality equation (5), which is in our opinion the deep expression of the notion of a gauge symmetry. In this sense, the enlarged space-time $\{x^\mu, \theta, \bar{\theta}\}$, although it may appear as a pure kinematical entity, contains in itself all ingredients that are relevant for introducing internal local symmetries, and suggests therefore by its own existence the necessity of internal symmetries.

An interesting, but as yet unexplored possibility, is that the constraints (5) that express the symmetry, and the equations of motion for the dynamics as deduced from the invariant Lagrangian (11), could be determined from a unique action principle in the enlarged space.

The following construction, although it contains a drawback due to an explicit θ, $\bar{\theta}$ dependence in the action that we shall obtain, supports such a possibility.*

Consider the θ and $\bar{\theta}$ expansion of each component in $\tilde{B}_p$

$$B^{q,\bar{q}}_{p-g}(x, \theta, \bar{\theta}) = B^{q,\bar{q}}_{p-g}(x, 0, 0) + \bar{\theta} s B^{q,\bar{q}}_{p-g}(x, 0, 0) + \theta \bar{s} B^{q,\bar{q}}_{p-g}(x, 0, 0) + \theta\bar{\theta} s\bar{s} B^{q,\bar{q}}_{p-g}(x, 0, 0) \tag{12}$$

For the sake of notational simplicity, it is useful to introduce the following graded scalar product, which acts on p-form gauge fields and their ghosts:

$$B^{q,\bar{q}}_l \otimes C^{q',\bar{q}'}_{l'} = \tfrac{1}{2}\delta_{ll'}\delta_{q\bar{q}'}\delta_{q'\bar{q}} \times \{B^{q,\bar{q}}_{[\mu_1\cdots\mu_l]} C^{q'\bar{q}'}_{[\mu_1\cdots\mu_{l'}]} + (^-)^{(q+\bar{q})(q'+\bar{q}')} C^{q',\bar{q}'}_{[\mu_1\cdots\mu_{l'}]} B^{q,\bar{q}}_{[\mu_1\cdots\mu_l]}\} \tag{13}$$

Knowing the expression of a field strength $\tilde{G} = \tilde{d}\tilde{B} + K(\tilde{B})$ as in equation (3) we define as an action the following quantity:

$$I_1 = \int d^D x \, d\theta \, d\bar{\theta} \, \bar{\theta}\theta \{\tilde{G}(x, \theta, \bar{\theta}) \otimes \tilde{G}(x, \theta, \bar{\theta})\} \tag{14}$$

* For previous attempts in the case of the Yang–Mills symmetry, see Ref. 6.

Using the definition of $\tilde{G}$, equation (3), and of the scalar product, equation (13), one gets trivially by integration over the Grassmann variables $\theta, \bar{\theta}$ that

$$I_1 = \int d^D x\{(G_{[\mu_1 \cdots \mu_{p+1}]})^2 + G^{1,0}_{[\mu_1 \cdots \mu_p]} G^{0,1}_{[\mu_1 \cdots \mu_p]} + G^{2,0}_{[\mu_1 \cdots \mu_{p-1}]} G^{0,2}_{[\mu_1 \cdots \mu_{p-1}]} + (G^{1,1}_{[\mu_1 \cdots \mu_{p-1}]})^2 + \cdots + (G_0^{p+1,0} G_0^{0,p+1})^2\} \tag{15}$$

By considering all quantities $B(x, 0, 0)$, $sB(x, 0, 0)$, $\bar{s}B(x, 0, 0)$ as independent fields, one then finds that the equations of motions of the fields sB and $\bar{s}B$, which stem from the action I_1, are identical to the BRST constraints (5). In other words, the geometrical BRST equations (6) and (7) can be derived from the equations of motion of the action (14).

These equations of motion do not specify, however, completely the action of the BRST symmetry on the inner ghosts, as seen from equation (7). To cure this problem, a simple possibility is to add to I_1 the following action, depending on the auxiliary field $\tilde{b}_{p-1}$:

$$I_2 = \int d^D x \, d\theta \, d\bar{\theta} \, \bar{\theta}\theta \{\tilde{b}_{p-1}(x, \theta, \bar{\theta}) + \tilde{d} \cdot \tilde{B}_p(x, \theta, \bar{\theta})\} \otimes \{\tilde{b}_{p-1}(x, \theta, \bar{\theta}) + \tilde{d} \cdot \tilde{B}_p(x, \theta, \bar{\theta}\}$$

with

$$\tilde{d} \cdot B_K^{q,\bar{q}} \equiv (\partial^{\mu_1} B^{q,\bar{q}}_{[\mu_1 \cdots \mu_K]} + sB^{q,\bar{q}+1}_{[\mu_2 \cdots \mu_K]} - \bar{s}B^{q+1,\bar{q}}_{[\mu_2 \cdots \mu_K]}) \, dx^{\mu_2} \wedge \cdots \wedge dx^{\mu_K} \tag{16}$$

One can indeed verify that the equations of motions of the fields $\tilde{b}$ in l_2, combined with these of l_1, are identical to the BRST equations (9), up to the mere change of variables $b^{q,\bar{q}}_{\mu_1 \cdots \mu_{p-1-g}} \to \frac{1}{2}(b^{q,\bar{q}}_{\mu_1 \cdots \mu_{p-1-g}} - \partial^{\mu_{p-g}} B_{[\mu_1 \cdots \mu_{p-g}]} + [K_{p+1}]^{q+1,\bar{q}+1}_{\mu_1 \cdots \mu_{p-g-1}})$. In the Yang-Mills theory, for instance, one has from the equations of motion of l_1

$$s\bar{c} + \bar{s}c = -[\bar{c}, c] \tag{17a}$$

and from the equations of motion of l_2

$$b + \partial^\mu A_\mu + s\bar{c} - \bar{s}c = 0 \tag{17b}$$

which allows one to determine separately $\bar{s}c$ and $s\bar{c}$. Notice the relevance of the minus sign in $s\bar{c} - \bar{s}c$, equation (17b). This sign comes from the definition (13) of the scalar product.

Observe finally that the action $I_1 + I_2$ is not gauge fixed since it reduces to the gauge invariant action

$$\int d^D x[G_{[\mu_1 \cdots \mu_{p+1}]}]^2 \tag{18}$$

after the determination of the BRST equations from the equation of motion of $s\tilde{B}$, $\bar{s}\tilde{B}$, and b. A gauge-fixing action is simply

$$I_3 = \int d^Dx\, d\theta\, d\bar{\theta}\, \tilde{B}_p(x, \theta, \bar{\theta}) \otimes \tilde{B}_p(x, \theta, \bar{\theta})$$

$$= \int d^Dx\, s\bar{s}[\tilde{B}_p(x, 0, 0) \otimes \tilde{B}_p(x, 0, 0)] \tag{19}$$

Indeed, the expansion of the $s\bar{s}$ term in the last equation, using the expression of s, $\bar{s}$, shows that I_3 provides a Landau-type gauge-fixing term for the action $I_1 + I_2$, with all relevant ghost interactions [3].

Collecting our results, we finally come to the conclusion that the following action

$$I = \int d^Dx\, d\theta\, d\bar{\theta}\{\bar{\theta}\theta[\tilde{G}_{p+1}(x, \theta, \bar{\theta}) \otimes \tilde{G}_{p+1}(x, \theta, \bar{\theta})$$

$$+ (\tilde{b}_{p-1} + \tilde{d} \cdot \tilde{B}_p) \otimes (\tilde{b}_{p-1} + \tilde{d} \cdot \tilde{B}_p)] + \lambda \tilde{B}_p \otimes \tilde{B}_p\} \tag{20}$$

seems to be an interesting starting point for a BRST invariant and gauge fixed field theory, only expressed in terms of the generalized classical-ghost objects $\tilde{B}$, $\tilde{b}$, and $\tilde{d}$. In this approach λ must be considered as an arbitrary gauge parameter. Similar results could be obtained for gravity and gauge theories coupled to gravity, using the techniques detailed in Ref. 2.

The existence of the action (20) suggests the possibility of a simple diagrammatic expansion for perturbative computations, in which many fewer diagrams would appear than in the ordinary component formalism, since closed loops of ghost and classical fields would be unified. However, the situation is not fully satisfying for at least two reasons: (i) The action (20) depends explicitly on θ and $\bar{\theta}$, in contrast to what is expected in a formalism in which x^μ, θ, $\bar{\theta}$ would be treated on an equal footing. (ii) We have been unable to determine a geometrical principle leading to the action (20).

REFERENCES

1. L. Baulieu, Cargese lecture 1983, LPTHE preprint 84.04, published in *Perspectives in Particles and Fields* (J. L. Basdevant and M. Lévy, eds.) Plenum Press, New York, 1983; Les Houches lectures, 1985, LPTHE preprint 85.43, to appear in *Architecture of Fundamental Interactions* (P. Ramond and R. Stora, eds.), North-Holland, Amsterdam, 1985.
2. L. Baulieu and M. Bellon, *Phys. Lett. 161B*, 96 (1985); *Nucl. Phys.* B*266*, 75 (1986); L. Baulieu, M. Bellon, and R. Grimm, *Nucl. Phys. B* (to be published).
3. L. Baulieu and L. Alvarez-Gaume, *Nucl. Phys. B212*, 255 (1983).
4. L. Baulieu and M. Henneaux, *Nucl. Phys. B277*, 268 (1985).

5. L. Baulieu, *Phys. Lett. 126B*, 455 (1983); L. Baulieu and J. Thierry-Mieg, *Phys. Lett. 144B*, 221 (1984).
6. S. Ferrara, O. Piguet, and M. Schweda, *Nucl. Phys. B119*, 493 (1977); K. Fujikawa, *Prog. Theor. Phys. 59*, 2045 (1978); *63*, 1364 (1980); L. Baulieu, unpublished, 1981; R. Delbourgo and P. D. Jarvis, *J. Phys. A. Math. Gen. 15*, 330 (1982); L. Bonora, P. Pasti, and M. Tonin, *Ann. Phys. 144*, 15 (1982); A. Hirshfield and H. Leschke, *Phys. Lett. 101B*, 48 (1981); J. Hoyos, M. Quiros, J. Ramirez Mittelbrunn, and F. J. de Urries, University of Madrid preprint, 1982.

Chapter 5

Geometry of String Space and String Field Theory

Ninoslav Bralić

1. INTRODUCTION

The impressive recent progress in string theory [1] has taken place for the most part at the level of the first quantized formalism, while the formulation in the language of field theory is far less understood. In a second quantized, field theoretic treatment, the fundamental object is the string field $\Phi\,[X]$, which is a functional of the string configuration $x^{\mu}(\sigma)$ becoming, upon quantization, an operator creating a string in that configuration. Formally, the classical theory will be defined by an action given in terms of a Feynman-type path integral of the form

$$S[\Phi] = \int \mathscr{D}x^{\mu}(\sigma)\mathscr{L}[\Phi, \delta\Phi/\delta x] \tag{1}$$

while the quantum theory is obtained from the generating functional

$$Z = \int \mathscr{D}\Phi[X]e^{iS} \tag{2}$$

While a field theory for bosonic strings was first formulated more than a decade ago [2], it was based on the light-cone-gauge quantization of the

Ninoslav Bralić • Facultad de Física, Pontificia Universidad Católica de Chile, Santiago 22, Chile.

string and did not provide therefore a covariant and gauge invariant formalism. Only recently a covariant gauge-fixed action has been constructed [3], and this has been followed by a rapid development of a gauge invariant free string theory [4, 5]. However, in their present form, the complexity of these formulations obscures the simple geometric contents that one expects the field theory to inherit from its first quantized form, and offers very little in terms of unveiling the underlying basic principle on which the theory relies.

Here we address these problems from a geometric point of view. Specifically, we characterize the configuration space $\mathscr{C}$ of bosonic strings as an infinite-dimensional manifold and explore the possibilities of implementing the required gauge invariance under string reparametrizations as isometries. Although we succeed in providing an explicit and unique gauge invariant geometric structure in $\mathscr{C}$, we show that this necessarily involves a degenerate, noninvertible metric in that manifold. This prevents us at this point from a straightforward construction of a string field theory in $\mathscr{C}$ on analogy with the conventional "point-field" case. Yet, our results do provide an insight as to the geometrical contents of the gauge invariance principle of the second quantized theory, and suggest the future steps needed to overcome the difficulties encountered at this stage.

2. LOCAL REPARAMETRIZATIONS

We restrict our attention to the case of free open bosonic strings in a space-time manifold M. Coordinates in M will be denoted by x^μ and we allow for a possibly nonflat background metric $g_{\mu\nu}$ on M. We denote by $\mathscr{C}$ the set of all parametrized strings in M, which can be specified by their space-time coordinate functions $x^\mu(\sigma)$, $\sigma \in [0, \pi]$, subject to the usual boundary conditions ($' = d/d\sigma$):

$$x'^\mu(\sigma)|_{\sigma=0,\pi} = 0 \tag{3}$$

Although we refer to the points in $\mathscr{C}$ as strings, it should be stressed that $\mathscr{C}$ is a space of parametrized objects ("loop space" in the case of closed strings). Thus, two strings that occupy the same space-time points in M, but that differ in their parametrization, are different points in $\mathscr{C}$. Since a string is uniquely specified by its space-time coordinate functions $x^{\mu(\sigma)}$, we shall refer to them as the space-time coordinates of the points in $\mathscr{C}$.

To specify the string field action beyond the formal expression in equation (1) we must impose appropriate invariance principles. The essential ingredient in string theory is the requirement of invariance under string reparametrizations. As viewed from $\mathscr{C}$ a (global) reparametrization f is a mapping given by

$$f\colon x^{\mu}(\sigma) \to x_f^{\mu}(\sigma) = x^{\mu}(f(\sigma)) \tag{4}$$

where $f(0) = 0$, $f(\pi) = \pi$, $f'(\sigma) > 0$. For infinitesimal reparametrizations we have

$$f(\sigma) = \sigma + h(\sigma), \qquad h(\sigma)|_{\sigma=0,\pi} = 0$$

and

$$x_f^{\mu}(\sigma) = x^{\mu}(\sigma) + h(\sigma)x'^{\mu}(\sigma)$$

On a scalar field $\Phi[X]$ defined on $\mathscr{C}$ this induces a transformation

$$\begin{aligned}\Phi_f[X] = \Phi[X_f] &= \Phi[X] + \int d\sigma\, h(\sigma)x'^{\mu}(\sigma)\frac{\delta\Phi}{\delta x^{\mu}(\sigma)}\\ &= \exp\left[i\int d\sigma\, h(\sigma)\mathscr{P}(\sigma)\right]\Phi[X]\end{aligned} \tag{5}$$

where

$$\mathscr{P}(\sigma) = \frac{1}{i}x'^{\mu}(\sigma)\frac{\delta}{\delta x^{\mu}(\sigma)} \tag{6}$$

Since in this case all strings are reparametrized in the same way, we refer to this type of transformation as *global* reparametrizations in $\mathscr{C}$.

Requiring the invariance of the string field action under these transformations leads, however, to a global invariance which is far smaller than the local gauge invariance of the first quantized theory. Indeed, is easy to check that the $\mathscr{P}(\sigma)$'s in equation (6) generate an infinite-dimensional Lie algebra characterized by

$$[L_h, L_{\tilde{h}}] = L_{(h'\tilde{h}-h\tilde{h}')}$$

where

$$L_h = \int d\sigma\, h(\sigma)\mathscr{P}(\sigma)$$

Transforming this to the Fourier normal modes of $x^{\mu}(\sigma)$, it is easy to show that this is the subalgebra of the full Virasoro algebra corresponding to σ reparametrizations. On the other hand, the first quantized theory is invariant under general reparametrizations in the τ–σ plane of the world surface $x^{\mu}(\tau, \sigma)$ describing the τ evolution of the string $x^{\mu}(\sigma)$. This corresponds to a local gauge invariance giving rise to the full Virasoro algebra, which, in turn, is the key ingredient in the elimination of ghosts from the theory [6]. In the second quantized theory there is no τ evolution parameter, but we deal with many string states and the global invariance under the reparametrization of all strings by the same transformation will not suffice to eliminate the ghosts. We need instead a local gauge invariance allowing the freedom of separate and independent reparametrizations of all strings entering in a physical process, either in external legs or in intermediate states. Only then will we be able to force each string into a light-cone-gauge condition, thus allowing the propagation of only physical degrees of freedom [8]. In the context of a field theory defined on $\mathscr{C}$, this means we must require the invariance under local reparametrization transformations in $\mathscr{C}$, in which the way a string is reparametrized depends on the string itself. Explicitly, a *local* reparametrization f will be a mapping in $\mathscr{C}$ given by

$$f: x^{\mu}(\sigma) \rightarrow x_f^{\mu}(\sigma) = x^{\mu}(f[X; \sigma]) \tag{7}$$

where now $f[X; \sigma]$ is a functional of the string as well as a function of σ. Similarly, for infinitesimal reparametrizations we have

$$f[X; \sigma] = \sigma + h[X; \sigma]$$

and

$$x_f^{\mu}(\sigma) = x^{\mu}(\sigma) + h[X; \sigma]x'^{\mu}(\sigma)$$

The transformation induced on a scalar field will now be

$$\begin{aligned}\Phi_f[x] &= \Phi[x] + \int d\sigma\, h[X; \sigma]x'^{\mu}(\sigma)\frac{\delta\Phi}{\delta x^{\mu}(\sigma)} \\ &= \exp\left\{i\int d\sigma\, h[X; \sigma]\mathscr{P}(\sigma)\right\}\Phi[x]\end{aligned} \tag{8}$$

This gauge transformation has the same geometrical contents as the so-called "chordal" gauge transformation introduced in Refs. 3 and 4. However, as is clear from the transformation law in equations (5) and (8), we are choosing here to implement the invariance of the theory on a "matter" field rather than on a "connection" field as in Refs. 3 and 4. The precise relation between the two approaches has not been worked out yet [8].

3. INVARIANT GEOMETRY IN STRING SPACE

The transformations under local reparametrizations in equations (7) and (8) can be viewed also as coordinate transformations in $\mathscr{C}$, and the implementation of the invariance under a given group of coordinate transformations is a standard geometric problem which can be approached with conventional tools. Consider first the simpler case of a conventional scalar field $\phi(x)$ defined on a space-time manifold M. A coordinate transformation in M induces the transformation

$$\phi_{(x)} \rightarrow e^{i\varepsilon^{\mu}(x)p_{\mu}}\phi_{(x)}$$

where $p_\mu = (1/i)\partial/\partial x^\mu$ and $\varepsilon^\mu(x)$ is the vector field on M generating the transformation. The allowed class of fields $\varepsilon^\mu(x)$ will span a Lie-bracket algebra corresponding to the group $\mathscr{G}$ of coordinate transformations being considered. If M can be endowed with a $\mathscr{G}$-invariant metric, so that $\mathscr{G}$ becomes the group of isometries of M, a $\mathscr{G}$-invariant kinetic term for the action of the field $\phi(x)$ will be given by the standard expression

$$S_0 = \frac{1}{2}\int dx\sqrt{\det g}\, g^{\mu\nu}\frac{\partial\phi}{\partial x^\mu}\frac{\partial\phi}{\partial x^\nu} \tag{9}$$

It is important to note that if the metric is not treated as a dynamical object but is kept as a given background, then the theory defined by S_0 will not be invariant under general coordinate transformations but only under the group of isometries of the metric (e.g., the Poincaré group in the case of the Minkowski metric).

Of course, it may not be possible to find an invariant metric for every group of transformations in M and, in a sense, we will face that difficulty in the string case. Yet, we postpone the discussion of that problem for later and proceed here to translate to the string field theory case this covariant approach. To that end we need to develop in $\mathscr{C}$ the conventional tools of differential geometry, of which we give here only a brief account limited to those aspects that are essential for our purpose. (For a detailed account see Ref. 7.) Specifically, we need to clarify the meaning of vector and tensor fields in $\mathscr{C}$ and determine their transformation laws under local reparametrizations.

To identify what tangent vectors in $\mathscr{C}$ are, consider a path $\xi(\tau)$ through $\mathscr{C}$, parametrized by τ, which at $\tau = 0$ starts from a given point $\xi(0)$ in $\mathscr{C}$. In space-time M this will be given by a two-dimensional surface $x^\mu(\tau, \sigma)$ describing the τ evolution of the initial string $x^\mu(\sigma) = x^\mu(\tau, \sigma)|_{\tau=0}$. The τ derivative of the path $\xi(\tau)$ at $\tau = 0$ will be a tangent vector in $\mathscr{C}$ at the initial point $\xi(0)$, which will be specified in M by the space-time functions $\dot{x}^\mu(0, \sigma)$, themselves the components of a tangent vector in M at the point $x^\mu(\sigma)$. Further, owing to the boundary conditions in equation (3), they satisfy $\dot{x}^\mu(0, \sigma)|_{\sigma=0,\pi} = 0$. Thus, in general, if V is a tangent vector field in $\mathscr{C}$, at a point in $\mathscr{C}$ with space-time coordinates $x^\mu(\sigma)$ it defines a mapping into tangent vectors in M at the points along the string, given by

$$V[X]\colon \sigma \to V^\mu[X; \sigma]$$

satisfying $V'^\mu[X; \sigma]|_{\sigma=0,\pi} = 0$. Here $V^\mu[X; \sigma]$ is a functional of the string as well as a function of σ, while μ is an M-vector index at the point $x^\mu(\sigma)$ in M. We shall refer to $V^\mu[X; \sigma]$ as the space-time components of the vector field V.

Similarly, a rank-two covariant vector field G in $\mathscr{C}$ can be specified by its space-time components $G_{\mu\nu}[X; \sigma, \sigma']$, which is a functional of the string with μ and ν being M-covariant indices at the points $x^\mu(\sigma)$ and $x^\nu(\sigma')$ in M. The contraction of two vector fields U and V with the tensor field G, at a point in $\mathscr{C}$ with space-time coordinates $x^\mu(\sigma)$, comes out as expected to be given by

$$G(U, V)|_X = \int d\sigma \int d\sigma' G_{\mu\nu}[X; \sigma, \sigma'] U^\mu[X; \sigma] V^\nu[X; \sigma']$$

which is a functional of the string on which is evaluated.

Although covariant expressions such as contractions are local in $\mathscr{C}$, we are interested in string theory which is local in space-time. Thus, we will consider only those tensors in $\mathscr{C}$ whose space-time components are of the form

$$G_{\mu\nu}[X; \sigma, \sigma'] = G_{\mu\nu}[X; \sigma]\delta(\sigma - \sigma') \tag{10}$$

so their contractions in $\mathscr{C}$ lead to local expressions in M. On the other hand, we are not interested in the intrinsic geometry of $\mathscr{C}$ on its own right, and therefore only those tensors that can be built from geometric objects in space-time will be of interest to us. This leaves us with only two candidates of physical interest, which we denote by $G^{(0)}$ and T, whose space-time components are given by

$$G^{(0)}_{\mu\nu}[X; \sigma] = g_{\mu\nu}(x(\sigma)) \tag{11}$$

where $g_{\mu\nu}$ is the background metric in M, and

$$T_{\mu\nu}[X; \sigma] = x'_\mu(\sigma) x'_\nu(\sigma) \tag{12}$$

(Either tensor could be multiplied by an arbitrary scalar $\psi[X; x'^\mu(\sigma)]$.) Tensors built from higher-order derivatives of $x^\mu(\sigma)$ will not be considered since presumably they cannot be regulated under the path integral in the action for the string field.

Under the action of the group of reparametrizations on $\mathscr{C}$ vectors and tensors transform covariantly so as to preserve contractions:

$$G_f(U_f, V_f)|_{X_f} = G(U, V)|_X$$

where the subindex f denotes the transformed objects under the reparametrization f. Thus, the statement of $\hat{G}$ being a reparametrization invariant tensor, namely, $\hat{G}_f = \hat{G}$, amounts to the invariance condition

$$\hat{G}(U_f, V_f)|_{X_f} = \hat{G}(U, V)|_X \tag{13}$$

For infinitesimal reparametrizations this leads to Killing's equation in $\mathscr{C}$ for the tensor $\hat{G}$. However, it is easier to solve equation (13) directly for finite transformations. To that end notice that if $V[X]$ is tangent to a path $\xi(\tau)$ through $\mathscr{C}$ at its initial point $\xi(0)$, then $V_f[X_f]$ is tangent to the path $\xi_f(\tau)$ at its initial point $\xi_f(0)$. Consider first the case of global reparametrizations. Then, the space-time coordinates $x_f^\mu(\tau,\sigma)$ and $x^\mu(\tau,\sigma)$ of the paths $\xi_f(\tau)$ and $\xi(\tau)$ are related by [cf. equation (4)]

$$x_f^\mu(\tau,\sigma) = x^\mu(\tau, f(\sigma))$$

and differentiating in τ at $\tau = 0$ we get

$$V_f^\mu[X_f;\,\sigma] = V^\mu[X;\,f(\sigma)] \tag{14}$$

Substituting this into equation (13) we find that the condition of invariance of $\hat{G}$ under *global* reparametrizations becomes

$$\hat{G}_{\mu\nu}[X_f;\,\sigma] = f'(\sigma)\hat{G}_{\mu\nu}[X;\,f(\sigma)] \tag{15}$$

In order to satisfy this condition the tensors $G^{(0)}$ and T in equations (11) and (12) must be modified to $\hat{G}^{(0)}$ and $\hat{T}$ given by

$$\hat{G}^{(0)}_{\mu\nu}[X;\,\sigma] = [x'^2(\sigma)]^{1/2} g_{\mu\nu}(x(\sigma)) \tag{16}$$

and

$$\hat{T}_{\mu\nu}[X;\,\sigma] = \frac{1}{[x'^2(\sigma)]^{1/2}}\, x'_\mu(\sigma) x'_\nu(\sigma) \tag{17}$$

which are now unique up to a reparametrization-invariant factor.

If we repeat this construction for a local reparametrization f, the space-time coordinates of the paths $\xi_f(\tau)$ and $\xi(\tau)$ through $\mathscr{C}$ will now be related by [cf. equation (7)]

$$x_f^\mu(\tau,\sigma) = x^\mu(\tau, f[X(\tau);\,\sigma]) \tag{18}$$

where $f[X(\tau);\,\sigma]$ is now a functional of the string being reparametrized and, therefore, is a function of τ. Explicitly,

$$\frac{d}{d\tau} f[X(\tau);\,\sigma] = \int d\sigma'\, \dot{x}^\mu(\tau,\sigma') \frac{\delta f[X(\tau);\,\sigma]}{\delta X^\mu(\tau,\sigma')}$$

Then, differentiating equation (18) in τ at $\tau = 0$, we now find for tangent vectors

$$V_f^\mu[X_f;\sigma] = V^\mu[X; f[X;\sigma]] + x'^\mu(f[X;\sigma]) \int d\sigma' \frac{\delta f[X;\sigma]}{\delta x^\nu(\sigma')} V^\nu[X;\sigma'] \tag{19}$$

Notice that this differs from the transformation law under global reparametrizations, equation (14), by a term that is nonlocal in M and that is independent of the vector V since the derivative $\delta f/\delta x^\mu$ is arbitrary.

Thus, substituting into equation (13) we now find that if $\hat{G}$ is to be invariant under *local* reparametrizations, it must be degenerate (noninvertible), being

$$\hat{G}_{\mu\nu}[X;\ \sigma]x'^{\nu}(\sigma) = 0 \tag{20}$$

On the other hand, since $\hat{G}$ must necessarily be a linear combination of the globally invariant tensors $\hat{G}^{(0)}$ and $\hat{T}$ in equations (16) and (17), we are forced to set

$$\hat{G}_{\mu\nu}[X;\ \sigma] = [x'^2(\sigma)]^{1/2}\left[g_{\mu\nu}(x(\sigma)) - \frac{x'_\mu(\sigma)x'_\nu(\sigma)}{x'^2(\sigma)}\right] \tag{21}$$

which is now unique (up to an overall reparametrization-invariant factor $\psi[X]$). It is now easy to check that $\hat{G}$ indeed satisfies the invariance condition in equation (13) for arbitrary local reparametrizations.

4. SUMMARY AND OUTLOOK

We have been able to identify a unique covariant tensor $\hat{G}$ in $\mathscr{C}$ which is invariant under local reparametrizations. However, we also found that $\hat{G}$ is necessarily degenerate, and this spoils the possibility of formulating a string field theory as a metric theory on $\mathscr{C}$ in analogy with the conventional point-field case in equation (9). The degeneracy of $\hat{G}$ means that, owing to the required invariance under local reparametrizations, one lacks in $\mathscr{C}$ the usual isomorphism between tangent and cotangent bundles (obtained by "raising" and "lowering" indices with an invertible tensor). Hence, it is important to notice that the object that enters directly in the construction of an action is a contravariant tensor with the appropriate invariance—that is, an invariant metric in the cotangent bundle, which need not be the inverse of a metric in the tangent bundle. Thus, if the covariant geometric approach is to succeed, our analysis must be carried on further to construct a contravariant tensor in $\mathscr{C}$ that is invariant under local reparametrizations. This, of course, will again be a degenerate tensor, but that will only imply the existence of zero modes in the action as it should be in a local gauge theory.

The lack of a nondegenerate invariant metric presents also a difficulty in the construction of the invariant integration measure needed to formulate the action of the thoery. This, however, is not a serious problem in the string case since, at least for free strings in a flat space-time, the invariant measure is known in terms of the vacuum wave functional of the string [4]. Still, a deeper understanding of the invariant measure in geometric terms would be desirable.

Another question which is left open is the relation between the matter field approach adopted here and the connection, or gauge field approach of other authors [3, 4].* This question, important in its own right, will be especially relevant in the incorporation of interactions, since in the gauge field point of view they are expected to arise from a non linear extension of the gauge principle. Finally, one should also extend this covariant approach to a geometric treatment of superstrings.

DISCUSSION

C. TEITELBOIM: You start with a scalar functional field. Is it obvious in your approach that the string field should be a scalar?

N. BRALIĆ: I don't think it is obvious from the start. But for the purpose of constructing an invariant geometry in string space the field is an auxiliary object and choosing it a scalar is not essential.

P. VAN NIEUWENHUIZEN: You have regarded reparametrizations as transformations moving one point to a different point in string space, and then as coordinate transformations of the same point. Are these equivalent as in general relativity?

BRALIĆ: Yes, they are. Here we have avoided the usual covariant expressions for coordinate transformations since they look rather singular in the "space-time" coordinate system in string space. But if you introduce well-behaved coordinates, such as the Fourier modes in the case of a flat space-time, the equivalence of the "active" and "passive" points of view become evident.

TEITELBOIM: Why should the usual Fourier mode expansion be restricted to a flat space-time?

BRALIĆ: It is not in the analytical sense. But it spoils covariance in a curved space-time since you separate the space-time vector index μ from the index σ labeling the point $x^{\mu}(\sigma)$ where that vector index transforms.

D. GROSS: String theory is a local finite theory because its interactions are given by the splitting and joining of strings at a point. But that would require a nonlocal interaction in string space $\mathscr{C}$. If that were the natural configuration space one would expect a local interaction in $\mathscr{C}$ of the form Φ^3. But that gives a nonlocal coupling of the world surfaces of the strings, which is incorrect.

VAN NIEUWENHUIZEN: But what would be wrong with a nonlocal interaction in string space?

GROSS: The point is that this covariant approach may be adequate for the kinetic piece of the action. But I just don't think it will be a natural point of view when it comes to handle the interactions.

* After this talk was delivered, the author learned of the work of K. Bardakci [8]. The starting point in that work is the same invariance under local reparametrizations considered here, and it provides some light as to the relation between a matter field and a connection field approach.

BRALIĆ: I agree that a geometric formulation along these lines can only determine the free part of the theory since it is clear that the interactions bear no relation to the local geometry of string space $\mathscr{C}$. But this does not mean that the interactions, which as you say will be nonlocal in $\mathscr{C}$, will violate the gauge symmetry of the free theory formulated in terms of the geometry of $\mathscr{C}$. On the other hand, from the point of view adopted here, the interactions will be related to other properties of $\mathscr{C}$ that we have not discussed—namely, its algebraic structure under the composition of parametrized strings. Clearly, that will lead to a local joining and splitting in space-time, but at this point I don't have much to say on how to implement this.

GROSS: Instead of generalizing the geometry of point fields to fields defined on the infinite-dimensional "loop space," one can proceed differently arriving at the noncommutative geometry proposed by E. Witten.

BRALIĆ: I am expecting to hear your account of Witten's work. It is not known to me yet.

REFERENCES

1. M. B. Green and J. H. Schwarz, *Phys. Lett. 149B*, 117 (1984); D. J. Gross, E. Martinec, J. A. Harvey, and R. Rohm, *Phys. Rev. Lett. 54*, 502 (1985). For a review see D. Gross, Chapter 3 in this volume.
2. M. Kaku and K. Kikkawa, *Phys. Rev. D10*, 1110, 1823 (1974).
3. W. Siegel, *Phys. Lett. 151B*, 391, 396 (1985).
4. T. Banks and M. E. Peskin, *Nucl. Phys. B264*, 513 (1986).
5. D. Friedan, *Nucl. Phys. B271*, 540 (1986); M. Kaku, *ibid. B267*, 125 (1986); A. Neveu and C. P. West, *Phys. Lett. 165B*, 63 (1985).
6. For a review see C. Rebbi, *Phys. Rep. 12C*, 1 (1974).
7. N. Bralić, *Prog. Theor. Phys. (Suppl.)86*, 93 (1986).
8. K. Bardakci, *Nucl. Phys. B271*, 561 (1986).

Chapter 6

Toward a Complete Theory for Unconventional Vacua

Mario Castagnino and Rafael Ferraro

1. INTRODUCTION

The vacuum is a well-defined notion in a free quantum field theory in unbounded flat space-time, if we use inertial observers; but it is an ill-defined notion if we try to work in a bounded or a curved space-time, or if we use accelerated observers. In these cases infinite new vacuum notion must be defined; and we must deal with unconventional vacua. In this chapter we introduce a reasonable vacuum definition in all the cases where we deal with noninertial observers in curved space-time (bounded space-time will be treated elsewhere).

To begin with we can remark that a frequently used procedure to define the vacuum is Hamiltonian minimization (or diagonalization). But all Hamiltonians introduced in the literature, until now, are observer independent. On the contrary, we know that the vacuum is an observer-dependent notion; the most famous and simple example is the existence of two different vacua in two-dimensional flat space-time related with the Minkowski and Rindler coordinates (cf., e.g., Ref. 1). Therefore it is reasonable to think

MARIO CASTAGNINO • Instituto de Astronomía y Física del Espacio, 1428 Buenos Aires, Argentina; and Instituto de Física de Rosario, 2000 Rosario, Argentina. RAFAEL FERRARO • Departamento de Matemáticas, Facultad de Ciencias Exactas y Naturales, Universidad de Buenos Aires, Ciudad de Universitaria, 1428 Buenos Aires, Argentina.

that an observer Hamiltonian can be defined in such a way that its minimization would yield an observer-dependent vacuum definition. The successes of state-independent Hamiltonian diagonalization can be explained because they are studied in a particular observer system and nobody has tried to extend this definition to more general cases. We shall define an observer-dependent Hamiltonian in a very natural way, and we shall see how it works. Hamiltonian diagonalization has been strongly criticized [2], but we believe the major objections are overcome. We refer the reader to Ref. 3 for the discussion of this problem. A more complete version of this work can be found in Ref. 4.

2. REFERENCE SYSTEMS

It is well known that there is a small confusion between a physical observers system and a geometrical coordinate system (or chart) in several papers. Of course they are two different concepts, e.g., in classical physics, an observer's system is a rigid frame and a clock, where we can use all kinds of charts, for instance, Cartesian or polar coordinates. In curved space-time we cannot use a rigid frame and the natural generalization of the observer's system will be a timelike fluid of observers, each one endowed with a clock [5], i.e., a set of timelike paths, each one with a different parameter, the "time" measured by the clock. This time is not necessarily the proper time; it is only an arbitrary continuous function of space-time. Of course we can describe this fluid of observers with any chart we like. We will find that physics is, in fact, observer dependent, but it is of course, chart independent.

We shall restrict ourselves to irrotational fluid; thus we can define a set of orthogonal timelike hypersurfaces to the fluid paths, and we can define a parameter T, on each surface, such that the equations $T = \text{const}$ would define the orthogonal hypersurfaces. We shall call this parameter a "natural time." Of course, there exists an infinite set of natural times. We can pass from one to another via a continuous function $T \to T' = T'(T)$. We shall see that physics is independent of the natural time we use; it is only dependent on the chosen observer's fluid. Of course, in general, natural time is different from proper time.

We can label each fluid world line by three real parameters x^1, x^2, x^3,* and we can call x^0 to the natural time T induced by the fluid of observers. Then x^0, x^1, x^2, x^3 is a chart and every event of space-time has its coordinates x^0, x^1, x^2, x^3—namely, the space coordinates of the fluid world line, where the event hapens, plus the natural time measured by the clock of this world

* In an n-dimensional space-time, $n - 1$ parameters.

line when the event happens. We shall call this chart an adapted chart. Of course we can use this adapted chart or an arbitrary chart; this will be physically irrelevant.

The fluid paths can be parametrized by the proper time, or by the natural time T; thus we can define two timelike vector, related with the fluid:

$$v^\mu = \frac{dx^\mu}{dT} \quad \text{and} \quad u^\mu = \frac{dx^\mu}{d\tau} \tag{1}$$

where $x^\mu(T)$ and $x^\mu(\tau)$ are the parametric equations of the fluid paths. u^μ is a unitary vector while v^μ is not unitary, in general. Both are tangent to the fluid world lines.

3. THE HAMILTONIAN

We shall work in an n-dimensional Riemannian manifold and we shall use the Misner–Thorne–Wheeler convention (–, –, –) [6]. We shall study a neutral scalar field with mass m and the field equation

$$(\Box + m^2 + \xi R)\phi = 0 \tag{2}$$

where R is the curvature scalar and ξ a coupling constant. Let $T_{\mu\nu}$ ($\mu, \nu = 0, 1, \ldots, n-1$) be the ordinary energy-momentum tensor of the field; we define our observer-dependent Hamiltonian, at an orthogonal surface to the fluid Σ, for the observer's system with vectors u and v:

$$\begin{aligned} H_\Sigma &= \int_\Sigma T_{\mu\nu} v^\mu d\Sigma^\nu \\ &= \int_\Sigma T_{\mu\nu} v^\mu u^\nu d\Sigma \end{aligned} \tag{3}$$

This is, of course, the most natural generalization of the Hamiltonian concept in our case. In fact, in flat space-time and in an inertial reference system it turns out to be the ordinary Hamiltonian. Moreover, if v^μ is a Killing vector field, it is easy to demonstrate that H_Σ is a conserved quantity, as the energy must be, if there is an invariant group, namely, the translation along the Killing field. Even if v^μ is not a Killing field, the following argument shows that H_Σ is the most accurate generalization of the energy. The Hamiltonian must be an integral like

$$H_\Sigma = \int_\Sigma P_\mu \, d\Sigma^\mu \tag{4}$$

where P^μ is a vector that must be built using the energy-momentum tensor $T_{\mu\nu}$ and the vectors related to fluid paths: v^μ and u^μ. Because v^μ and u^μ

are proportional, the most general expression for P_μ is

$$P_\mu = T_{\mu\nu} v^\nu F(v) \tag{5}$$

where $F(v)$ is an arbitrary scalar function of $v \equiv v_\mu v^\mu$. But H_Σ must be somehow the canonical coordinate conjugated to T. Therefore, if we change the natural time T:

$$T \to T' = T'(T) \tag{6}$$

then H_Σ must change as

$$H_\Sigma \to H'_\Sigma = \frac{dT}{dT'} H_\Sigma \tag{7}$$

But H'_Σ is

$$\int_\Sigma P'_\mu \, d\Sigma^\mu = \int_\Sigma T_{\mu\nu} v'^\nu F(v') \, d\Sigma^\mu = \int_\Sigma T_{\mu\nu} \frac{dT}{dT'} v^\nu F(v') \, d\Sigma^\mu \tag{8}$$

Therefore, in order to satisfy equation (7)

$$F(v') = F(v) = \text{const} \tag{9}$$

and we may take this constant equal to unity. This proves that equation (3) is the good Hamiltonian. In the adapted chart the Hamiltonian reads

$$H_\Sigma = \int_\Sigma T_{00} g_{00}^{-1/2} \, d\Sigma \tag{10}$$

and the D'Alembert operator of equation (2):

$$\Box \phi = \frac{\phi_{,00}}{g_{00}} + \frac{\gamma^{ij}\gamma_{ij,0}}{2g_{00}} \phi_{,0} - \frac{\gamma^{ij} g_{00,j}}{2g_{00}} - \Delta\phi \tag{11}$$

where γ_{ij} is the spatial $(i, j = 1, 2, \ldots, n-1)$ metric on the hypersurface Σ and $\Delta\phi = \gamma^{ij}\phi_{,,ij}$ the Laplace operator on Σ. Using these equations, by partial integration one gets

$$H_\Sigma = \int_\Sigma d\Sigma \, g_{00}^{-1/2} \left[\frac{1}{2} (\phi_{,0})^2 - \frac{1}{4} (\phi\phi_{,00} + \phi_{,00}\phi) + \frac{B}{2} (\phi\phi_{,0} + \phi_{,0}\phi) + c\phi^2 \right] \tag{12}$$

where coefficients B and C are

$$B = \left(\xi - \frac{1}{4} \right) \gamma^{ij}\gamma_{ij,0} + \frac{g_{00,0}}{4g_{00}} \tag{13}$$

$$C = \xi \left[\frac{1}{4} \gamma^{kl}\gamma^{ij}\gamma_{ik,0}\gamma_{lj,0} + \frac{1}{4g_{00}} \gamma_{ij}\gamma_{ij,0} g_{00,0} - \frac{1}{2} \gamma^{ij}\gamma_{ij,00} \right] \tag{14}$$

This expression will be most useful in the computation. Let us now briefly review why the vacuum notion is ambiguous in the case we are dealing with. To quantize the theory, an orthonormal basis in the inner product

$$\langle \phi, \psi \rangle = i \int_{\Sigma} (\phi \overleftrightarrow{\partial}_{\mu} \psi) \, d\Sigma^{\mu} \tag{15}$$

of solutions of the field equation (2) must be used. Let us call $\{u_k\} \cup \{u_k^*\}$ to this basis, where k symbolizes the set of indices necessary to label each vector of the bases. Then we can expand the field operator as

$$\phi(x) = \sum_k a_k u_k(x) + a_k^{\dagger} u_k^*(x) \tag{16}$$

where a_k and a_k^* are annihilation and creation operators, and we can define the vacuum as the quantum state that is annihilated by all the a_k's

$$a_k|0\rangle = 0, \qquad \forall a_k \tag{17}$$

But if we change the basis to another one $\{\bar{u}_k\} \cup \{\bar{u}_k^*\}$ we can repeat the procedure, and in general, we shall find another vacuum $|\bar{0}\rangle \neq |0\rangle$.

In Minkowski unbounded space-time, if we work with inertial observers, we have a criterion to choose the right basis—Poincaré invariance—and we obtain the ordinary plane wave basis. But this invariance is broken if we use accelerated observers, or by the boundaries, if we study the case of bounded space-time, and it does not exist in curved space-time. Thus in all these cases we do not have a criterion to choose the basis, and the vacuum notion turns out to be ambiguous.

We propose the following definition to solve the problem. In an observer's system with vectors u^{μ} and v^{μ} and at an orthonormal surface Σ, the vacuum $|0, \Sigma\rangle$ will be the quantum state that minimizes the VEV of H_{Σ} (3), i.e.,

$$\langle 0, \Sigma | H_{\Sigma} | 0, \Sigma \rangle \tag{18}$$

If we consider that H_{Σ} is the natural generalization for the energy, this definition is also quite natural; $|0, \Sigma\rangle$ is the state of minimum energy. Surprisingly enough, this simple definition encompasses a great number of other definitions that can be found in the literature and are based in completely different ideas: particle detectors, conformal mapping, analytical continuation, etc.

Of course for another observer's system, or another surface Σ', we can have another vacuum: $|0, \Sigma'\rangle$. If

$$|0, \Sigma\rangle \neq |0, \Sigma'\rangle \tag{19}$$

there is a difference in the particle number between the two vacua, i.e., the observers of vacuum $|0, \Sigma\rangle$ will consider $|0, \Sigma'\rangle$ a many-particle system and vice versa.

In the next section we shall consider some cases where we can implement our definition.

4. THE CASE OF VARIABLE SEPARATION AND EXAMPLES

In the adapted chart, field equation (2) reads:

$$\frac{\phi_{,00}}{g_{00}} + \frac{\gamma^{ij}\gamma_{ij,0}}{2g_{00}}\phi_{,0} - \frac{\gamma^{ij}g_{00,j}}{2g_{00}}\phi_{,i} - A\phi + (m^2 + \xi R)\phi = 0 \tag{20}$$

Then we can implement our definition in the particular case where we can solve this equation by variable separation, i.e., a solution can be found with the form

$$\mu_k(x) = \Theta_k(x^0)\Sigma_k(x^i) \tag{21}$$

This can be done, for instance, in the following particular cases:

1. $R = R(x^0), \qquad g_{00} = g_{00}(x^0), \qquad \gamma_{ij} = f(x^0)\Delta_{ij}(x^k)$

   (i.e., if the curvature scalar and the component of the metric tensor turn out to be functions of only x^0, and γ_{ij} can be factorized as $f(x^0)\Delta_{ij}(x^k)$];
2. $R = R(x^k), \qquad g_{00} = g(x^0)h(x^i), \qquad \gamma_{ij} = \Delta_{ij}(x^k)$;
3. $m = 0, \qquad R = 0, \qquad g_{00} = g(x^0)h(x^i), \qquad \gamma_{ij} = f(x^0)\Delta_{ij}(x^k)$;
4. $g_{\mu\nu} = g_{\mu\nu}(x^0)$; in this case $\Sigma_k(x^i)$ will be $e^{i\mathbf{k}\cdot\mathbf{a}}$.

In all these cases coefficients B and C of equations (13) and (14) turn out to be constant on each orthogonal surface Σ, and if we normalize the spatial factor $\Sigma_k(x^i)$ as

$$\int_\Sigma |\Sigma_k|^2 g_{00}^{-1/2}\, d\Sigma = 1 \tag{22}$$

the equation for the time factor $\Theta_k(x^0)$ is

$$\Theta_{k,00} + b(x^0)\Theta_{k,0} + c_k(x^0)\Theta_k = 0 \tag{23}$$

where the coefficients $b(x^0)$ and $c_k(x^0)$ can be computed case by case. The Hamiltonian H_Σ then reads

$$\begin{aligned} H_\Sigma = \sum_k a_k a_k \{\tfrac{1}{2}(\Theta_{k,0})^2 - \Theta_k\Theta_{k,00}] \\ + B\Theta_k\Theta_{k,00} + c\Theta_k^2\}\left(\int_\Sigma \Sigma_k^2 g_{00}^{-1/2}\, d\Sigma\right) + a_k^\dagger a_k[\tfrac{1}{2}|\Theta_{k,0}|^2 - \tfrac{1}{4}(\Theta_k\Theta^*_{k,00} \\ + \Theta_{k,00}\Theta_k^*) + (B/2)(\Theta_k\Theta^*_{k,0} + \Theta_{k,0}\Theta_k^k) + c|\Theta_k|^2] + \text{H.c.} \end{aligned} \tag{24}$$

Where Θ_k, $\Theta_{k,0}$, and $\Theta_{k,00}$ are evaluated at Σ. The different basis, at a orthonormal surface Σ, can be obtained from their Cauchy data $\Theta_k|_\Sigma$ and $\Theta_{k,0}|_\Sigma$. These data are not independent, if the basis should be an orthonormal one, and they must satisfy the condition

$$(\Theta_k\Theta^*_{k,0} - \Theta^*_k\Theta_{k,0})|_\Sigma = i \tag{25}$$

Now we can compute the Cauchy data of the basis with a minimal VEV of the H_Σ if we compute the minimum of $\langle 0, \Sigma|H_\Sigma|0, \Sigma\rangle$, where H_Σ is taken from equation (24), for $\Theta_k|_\Sigma$ and $\Theta_{k,0}|_\Sigma$ variables, with the constraint (25). $\Theta_{k,00}|_\Sigma$ is a function of $\Theta_k|_\Sigma$ and $\Theta_{k,0}|_\Sigma$, from equation (23). The result is that the ratio of $\Theta_{k,0}$ and Θ_k must be

$$\left.\frac{\Theta_{k,0}}{\Theta_k}\right|_\Sigma = -\left(B + \frac{b}{2}\right) - i\left[2C + c_k + \left(B + \frac{b}{2}\right)^2\right]^{1/2} \tag{26}$$

From this equation and (25) we can obtain the Cauchy data that correspond to $|0, \Sigma\rangle$. The Hamiltonian in this basis reads

$$H_\Sigma = \sum_k \frac{1}{2}\left[2C + c_k - \left(B + \frac{b}{2}\right)^2\right]\Bigg|_\Sigma^{1/2} (a^\Sigma_k a^{+\Sigma}_k + a^{+\Sigma}_k a^\Sigma_k) \tag{27}$$

where a^Σ_k and $a^{+\Sigma}_k$ are the operators that correspond to vacuum $|0, \Sigma\rangle$. Then the vacuum that minimizes the Hamiltonian also diagonalizes it. In fact, it is easy to show that if we define the vacuum.

Let us see some examples of this kind of vacuum.

4.1. The Killing Vacua

If v^μ is a Killing vector field at least in a neighborhood of surface Σ, then in this neighborhood the coefficients read

$$B = 0, \qquad C = 0, \qquad b = 0; \qquad c_k = \omega^2_k \tag{28}$$

where ω^2_k is now a constant, and the basis functions are

$$\mu_k(x) = (2\omega_k)^{-1/2}\Sigma_k(x^i)e^{-i\omega_k x^0} \tag{29}$$

Thus in this case we obtain the characteristic factor $e^{-i\omega_k x^0}$ and we can interpret ω_k as the energy of the field in mode k. The Hamiltonian turns out to be

$$H_\Sigma = \tfrac{1}{2}\sum_k \omega_k(a^\Sigma_k a^{+\Sigma}_k + a^{+\Sigma}_k a^\Sigma_k) \tag{30}$$

The Minkowski and Rindler vacua, explained in the Introduction, are Killing vacua that correspond to the Killing vector field of straight lines parallel to the t axis, for the Minkowski vacuum, and to the Killing field of tangent vectors to the Rindler's hyperbolae, for the Rindler one (cf. Ref. 1). Also for every static metric we have a Killing vector field; then every static vacuum is a Killing vacuum—for instance, the static vacuum in the Einstein universe [7], or the Boulware vacuum for the static coordinates of Schwarzschild geometry. Moreover, de Sitter space has four timelike independent Killing vector fields, each one with their vacuum (cf. Refs. 8 and 9) etc. All Killing vacua are, in general, good vacua, in the sense that the particle difference between two Killing vacua turns out to be finite, and the renormalized VEV of the energy-momentum tensor is also finite. In fact, they seem to be strong vacua in the sense of Ref. 3. They are, of course, the first and more natural generalization of the ordinary Minkowski vacuum.

4.2. Comoving Observer's Fluid in Robertson–Walker Universe

We can use as observer's system the geodesic, comoving fluid of a Robertson-Walker universe with metric

$$ds^2 = dt^2 - a^2(t)[d\chi^2 + f^2(\chi)(d\theta + \sin^2\theta\, d\phi] \tag{31}$$

where

$$f(\chi) = \begin{cases} \sin\chi, & 0 \le \chi \le 2\pi, \quad \text{for spatially closed universe } (k = 1) \\ \chi, & 0 \le \chi \le \infty, \quad \text{for spatially flat universe } (k = 0) \\ \sinh\chi, & 0 \le \chi \le \infty, \quad \text{for spatially hyperbolic universe } (k = -1) \end{cases} \tag{32}$$

In this case the coefficients read

$$\begin{aligned} B &= 6\left(-\frac{1}{4}\right)\frac{\dot a}{a}, \qquad C = 3\xi\frac{\ddot a}{a} \\ b &= \frac{3\dot a}{a}, \qquad c_k = m^2 + \frac{k^2}{a^2} + 6\frac{\xi}{a^2}(a\ddot a + \dot a^2 + K) \end{aligned} \tag{33}$$

the minimization condition is

$$\frac{\Theta_{k,0}}{\Theta_k}\Big|_\Sigma = -6\xi\frac{\dot a}{a} - i\left[m^2 + \frac{k^2 + 6\xi K}{a^2} + \left(\frac{\dot a}{a}\right)^2 6\xi(1-6\xi)\right]^{1/2} \tag{34}$$

and the Hamiltonian turns out to be

$$H_\Sigma = \tfrac{1}{2}\sum_k [m^2 + k^2 + 6\xi(1-6\xi)]^{1/2}(a_k^\Sigma a_k^{+\Sigma} + a_k^{+\Sigma} a_k^\Sigma) \tag{35}$$

All the "Russian school" uses this prescription to define their vacua (cf., e.g., Ref. 10) in the Robertson-Walker universe. Also other vacua like the one obtained by the Wick trick by De Witt, the one introduced by Charach and Parker [11] via an analytic continuation, and the vacuum of Chitre and Hartle [12] in a linearly expanding universe, based on the

Feynman path integral, are of this kind (cf. Refs. 13, 14, and 3). Also the Hamiltonian minimization works in more general anisotropic universes like the Bianchi type I universe [3]. But in general, all these vacua are not as good as the Killing vacua. In fact, the particle difference between two of these vacua could be infinite, an obvious unphysical result. Thus, a new condition must be introduced, in addition to Hamiltonian diagonalization. For instance the VEV of the renormalized energy-momentum tensor must be finite [15], or the basis Cauchy data must coincide with the adiabatic ones up to the first adiabatic order [16]. (Then this vacuum becomes minimal or weak in the sense of Refs. 3 and 16.) It can be shown that if some vacua exist such that they satisfy both conditions on certain surfaces, then all the physics works correctly among these vacua.

5. THE CONFORMAL CASE AND EXAMPLES

Even if we cannot separate variables, we can implement our definition, using a conformal mapping. Suppose we have two manifolds, each one with an observer's system such that the metrics in the corresponding adapted charts would be related by

$$\bar{g}_{\mu\nu} = \Omega^2(x) g_{\mu\nu} \tag{36}$$

Let us study the case $m = 0$ and define the following mapping for the fields:

$$\bar{\phi}(x) = \Omega(x)^{(2-m)/2} \phi(x) \tag{37}$$

Then it is easy to show that the Hamiltonian is related by the following equation:

$$\begin{aligned}\bar{H}_\Sigma[\bar{\phi}] = H_\Sigma[\phi] + \left[\xi(m-1) - \frac{(m-2)}{4}\right] \\ \times \int d\Sigma\, g_{00}^{-1/2}\left[\frac{\Omega_{,0}}{\Omega}(\phi,\phi_{,0} + \phi_{,0}\phi)\right] \\ - \left[\frac{\Omega_{,00}}{\Omega} + (m-3)\left(\frac{\Omega_{,0}}{\Omega}\right)^2 + \frac{1}{2}\left(\frac{\Omega_{,0}}{\Omega}\right)\left(\gamma_{ij,0}\gamma^{ij} - \frac{g_{00,0}}{g_{00}}\right)\right]\phi^2\end{aligned} \tag{38}$$

Then if the coupling constant ξ is

$$\xi = \frac{m-2}{4(m-1)} \tag{39}$$

we have

$$\bar{H}_\Sigma[\bar{\phi}] = H_\Sigma[\phi] \tag{40}$$

We know that when ξ is given by equation (39), we have a conformal mapping of the theory, from one manifold to the other (cf. Ref. 1). Then we see that, in this case, if a vacuum and a basis minimize the Hamiltonian in one manifold, the basis obtained by the conformal mapping and the corresponding vacuum minimize the Hamiltonian in the other manifold. Thus if we can minimize the Hamiltonian in one manifold, say because we can make variable separation in this manifold, we can minimize the Hamiltonian in the other manifold where variable separation cannot be performed. Normally the first manifold is a Minkowski space or a static space-time, and the second one is related by a conformal transformation with the first one.

The vacuum of Norma Sánchez belongs to this class [17], and also those related with the nine conformal Killing fields of Minkowski space defined by Brown, Otewill, and Siklos [18]. Another example is the Kruskal vacuum in two-dimensional eternal black holes (cf. Ref. 1) and the vacuum of Ref. 19. All these vacua seem to be good vacua in the sense that particles difference among them and with the Killing vacua are finite (they are strong vacua in the sense of Refs. 3 and 16; thus no extra local condition is necessary in this case).

6. CONCLUSIONS

We believe this chapter shows the role played by the observers in the definition of a quantum vacuum. The main function of the observer's system is to define unambiguously a time notion—the natural time—and a space notion, through the foliation of the space-time by the hypersurfaces, orthogonal to the fluid world lines. The vacuum definition seems to be related with this decomposition of space-time into time and space components. Then the use of an irrotational fluid is essential.

The device to achieve our definition is our Hamiltonian, an observer-dependent, chart-independent operator. Can we extend our definition to encompass all the examples of unconventional vacua of the literature? To do so there are two kinds of problems:

- Mathematical problems: The mathematics becomes much more complicated when we study more realistic examples, like the black hole vacua. Normally, the field equation cannot be solved by closed procedures. This fact make the analysis much more difficult.
- Physical problems: Some vacua are related with null Killing vector field, like the Unruh vacuum in black holes. Our observer's fluid is a timelike one; thus we must either define a null observer's fluid, a system with a difficult physical interpretation, or else we must study

the null Killing vector field as a limit of the timelike fluid. On the other hand, a local extra condition must be generally added to obtain good vacua. This problem has been studied in Robertson–Walker and Bianchi type *I* universes, but not in the general case.

Neither the first nor the second class of problems seems unsolvable; thus perhaps we can display a complete theory for unconventional vacua in the near future.

DISCUSSION

C. ARAGONE: Is the theory independent of coordinate changes on the surface Σ? For instance, if surface Σ would have a symmetry group, would the vacuum notion be an invariant of the group?

M. CASTAGNINO: Yes, because this change is only a geometrical chart change that leaves the observer's fluid invariant. And the vacuum only depends on the fluid.

REFERENCES

1. N. D. Birrel and P. C. W. Davies, *Quantum Fields in Curved Space*, Cambridge University Press, Cambridge, 1982.
2. S. A. Fulling, *Gen. Rel. Grav. 10*, 807 (1979).
3. M. Castagnino and F. D. Mazzitelli, *Phys. Rev. D 31*, 742 (1985).
4. M. Castagnino and R. Ferraro, *Phys. Rev. D 34*, 497 (1986).
5. C. Cattaneo, *Nuovo Cimento 10*, 318 (1958).
6. C. W. Misner, K. S. Thorne, and J. A. Wheeler, *Gravitation*, Freeman, San Francisco, 1973.
7. R. Ruffini and S. Bonazzola, *Phys. Rev. 187*, 1767 (1969).
8. M. Castagnino, *C. R. Acad. Sci. Paris 268*, 1157 (1969).
9. N. A. Chernikov and B. A. Tagirov, *Ann. Inst. H. Poincaré 9*, 109 (1968).
10. A. A. Grib, S. G. Mamayev, and V. M. Mostepanenko, *J. Phys. A 13*, 2057 (1980).
11. Ch. Charach and L. Parker, *Phys. Rev. D 24*, 3023 (1981).
12. D. N. Chitre and J. B. Hartle, *Phys. Rev. D 16*, 251 (1977).
13. E. Calzetta and M. Castagnino, *Phys. Rev. D 28*, 1298 (1983).
14. E. Calzetta and M. Castagnino, *Phys. Rev. D 29*, 1609 (1984).
15. M. Castagnino, On the vacuum definition in curved space-time, in *Proceedings of the III Quantum Gravity Seminar, Moscow*, World Scientific, Singapore, 1985, p. 496.
16. M. Castagnino and L. Chimento, Weak and strong quantum vacua II, submitted to *Phys. Rev. D 34*, 3676 (1986).
17. N. Sánchez, *Phys. Rev. D 24*, 2100 (1981).
18. M. R. Brown, A. C. Otewill, and S. T. C. Siklos, *Phys. Rev. D 26*, 1881 (1982).
19. M. Castagnino and R. Ferraro, *Ann. Phys. (N.Y.) 161*, 1 1985.

Chapter 7

Some Properties of the Solutions of the Back-Reaction Problem

Mario Castagnino and Juan Pablo Paz

1. INTRODUCTION

The aim of this chapter is to present some qualitative considerations about the type of solutions of the semiclassical Einstein equations (s.E.e.). These equations [1], which rule the dynamic of the space-time metric, have a source term that is the renormalized expectation value of the energy-momentum tensor. The computation of this last object as a function of the metric is a rather difficult task that has been performed only in a few cases. The s.E.e. can be written as

$$R_{\mu\nu} - \tfrac{1}{2}Rg_{\mu\nu} + \Lambda g_{\mu\nu} + a_1\,{}^{(1)}H_{\mu\nu} + a_2\,{}^{(2)}H_{\mu\nu} = \kappa\langle\psi|T_{\mu\nu}|\psi\rangle_{\text{ren}}$$

$$^{(1)}H_{\mu\nu} = \frac{\delta}{\delta g^{\mu\nu}}\int d^4x(-g)^{1/2}R^2, \tag{1}$$

$$^{(2)}H_{\mu\nu} = \frac{1}{(-g)^{1/2}}\frac{\delta}{\delta g^{\mu\nu}}\int (-g)^{1/2}\,d^4x\,R_{\mu\nu}R^{\mu\nu}$$

Here, using an approach developed by Wada and Azuma [1], we will present all the different families of solutions of the equation (1) that occur

---

MARIO CASTAGNINO • Instituto de Astronomía y Física del Espacio, 1428 Buenos Aires, Argentina; and Instituto de Física de Rosario, 2000 Rosario, Argentina. JUAN PABLO PAZ • Instituto de Astronomía y Física del Espacio, 1428 Buenos Aires, Argentina.

when the field theory treated is classically conformally invariant. We discuss what modifications are introduced by the presence of massive fields. In this case particle creation takes place and the problem is more complicated.

The chapter is organized as follows: in Section 2 we review the method of Wada and Azuma, showing all the different families of solutions of equation (1) when the $\langle I_{\mu\nu}\rangle_{\text{ren}}$ is given by (2) (see below). In Section 3 we discuss the qualitative changes originated by the presence of massive fields. We discuss also the importance and the cosmological relevance of the ambiguity in the vacuum definition.

2. CONFORMALLY INVARIANT FIELDS

If the fields that are present are free and massless and they are conformally coupled to the scalar curvature, the renormalized expectation value of the $T_{\mu\nu}$ can be written as

$$
\begin{aligned}
T^1_{\mu\nu} &\equiv \langle 0|T_{\mu\nu}|0\rangle_{\text{ren}} \\
&= \frac{b_1}{2880\pi^2}\left(R_{\mu\rho}R^{\rho}_{\nu} - \tfrac{2}{3}RR_{\mu\nu} - \tfrac{1}{2}g_{\mu\nu}R_{\rho\theta}R^{\rho\theta} + \tfrac{1}{4}g_{\mu\nu}R^2\right) \\
&\quad + \frac{b_2}{17280\pi^2}\left(2R_{;\mu;\nu} - 2g_{\mu\nu}\Box R - 2RR_{\mu\nu} + \tfrac{1}{2}g_{\mu\nu}R^2\right) \qquad (2)
\end{aligned}
$$

(we restrict ourselves to the case where $C_{\mu\nu\rho\lambda} = 0$). In this expression the quantum state $|0\rangle$ is the known conformal vacuum and the coefficients b_i are given as combinations of N_i, the number of fields with spin i ($i = 0, \frac{1}{2}, 1$) by

$$b_1 = N_0 + \tfrac{11}{2}N_{1/2} + 62N_1, \qquad b_2 = N_0 + 6N_{1/2} - 18N_1$$

If we write down the (0, 0) component of equation (1) in Robertson-Walker spatially flat metrics the Hubble "constant" $H = \dot{a}/a$ satisfies the following differential equation:

$$
\begin{aligned}
H\ddot{H} &= \frac{\beta}{2\beta'}H^4 - \frac{1}{2\beta'}H^2 + \frac{1}{2}(\dot{H}^2 - 6\dot{H}H^2) + \frac{\Lambda}{6\beta'} \\
\beta &= \frac{\kappa b_1}{2880\pi^2}, \qquad \beta' = -\left(6a_1 + 2a_2 + \frac{\kappa b_2}{2880\pi^2}\right)
\end{aligned} \qquad (3)
$$

The way in which all the different families of solutions of this equation can be seen is by drawing trajectories in the phase space (i.e., $\dot{H}$ vs. H). In order to do that it is convenient to point out a few things.

Specializing $H = 0$ in (3) we can see that this value is only consistent with $\dot{H}_{\pm} = \pm(-\Lambda/3\beta')^{1/2}$. The obvious conclusion is that if $\Lambda/\beta' > 0$ the trajectories never cross the axis $H = 0$ and that if $\Lambda/\beta' < 0$ the trajectories cross the axis $H = 0$ only through the point $(0, \dot{H}_\pm)$. In the particular case where $\Lambda = 0$ we have $\dot{H}_\pm = 0$ and the trajectories cross the axis $H = 0$ through the origin. We immediately conclude that in this case the origin is an unstable solution in the sense that oscillations around this point (the Minkowski space) do not exist. This result is not clear in the literature and there are a few recent papers that assume that the Minkowski space is a stable solution of (3) [2] (this conclusion is obtained doing an incorrect linear analysis of this differential equation).

The existence of fixed points in the plane $(H, \dot{H})$ can be examined looking for solutions of (3) that satisfy $\dot{H} = 0 = \ddot{H}$. Those static solutions are the roots of the quadratic equation:

$$H^4 - \frac{1}{\beta} H^2 + \frac{\Lambda}{3} = 0$$

The existence and the value of the fixed points will depend on the value of the coefficient β and $D = 1 - 4\Lambda\beta/3$. We can immediately see that there are three different situations: if $D < 0$ there are no fixed points; if $D > 1$ there are two fixed points; if $0 < D < 1$ there are four fixed points if $\beta > 0$ and none if $\beta < 0$. The form of the trajectories near the fixed points can be found in each case by doing a linear analysis. In this way we can write linear equations for the "normal modes" that have solutions that are proportional to $\exp(\Lambda_\pm t)$, where $\Lambda_\pm$ are the eigenvalues of a linear operator. We can also compute the eigenvectors (the direction of the normal modes). Doing the detailed computation we can obtain that near a fixed point x_0 the values of $\Lambda_\pm$ are

$$\Lambda_\pm = \frac{B}{2} \pm \left(\frac{B^2}{4} + A \right)^{1/2}$$

where

$$B = -3x_0 \quad \text{and} \quad A = \left(X_0^2 - \frac{2\Lambda}{3} \right) \frac{1}{\beta' X_0^2}$$

Using this result we can draw all the possible trajectories that can be seen in Figs. 1–8 (see Ref. 1).

3. MASSIVE FIELDS

If we want to study the solutions of (1) when massive fields are present we should know the expression of $\langle T_{\mu\nu} \rangle_{\text{ren}}$ as some function of the metric (and its derivatives). In general this is not known. However, we can examine the way in which the different behaviors we mention in Section 2 change.

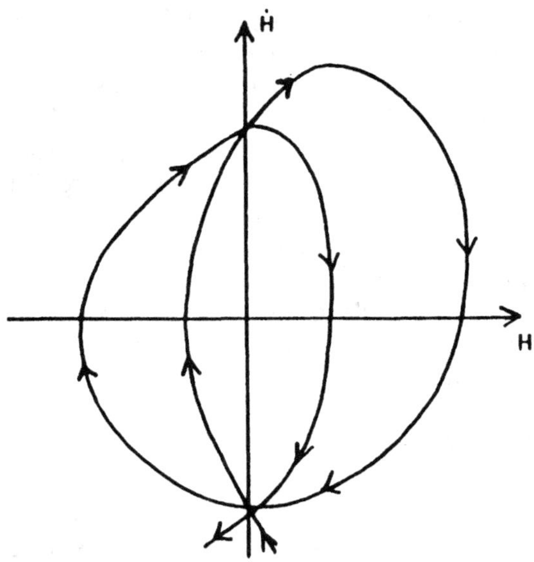

Figure 1. Trajectories without fixed points ($D < 0, \beta > 0, \Lambda > 0, \beta' < 0$).

The existence of fixed points (static solutions) can be studied in an analytic way, keeping in mind that we know the expression for $\langle T_{\mu\nu}\rangle_{\text{ren}}$ in the de Sitter universe for massive fields of arbitrary spin. Using the result of previous papers (see Ref. 3 and references therein) we can write in the case where N_i fields of spin i with mass m_i are present ($i = 0, \frac{1}{2}$) the following

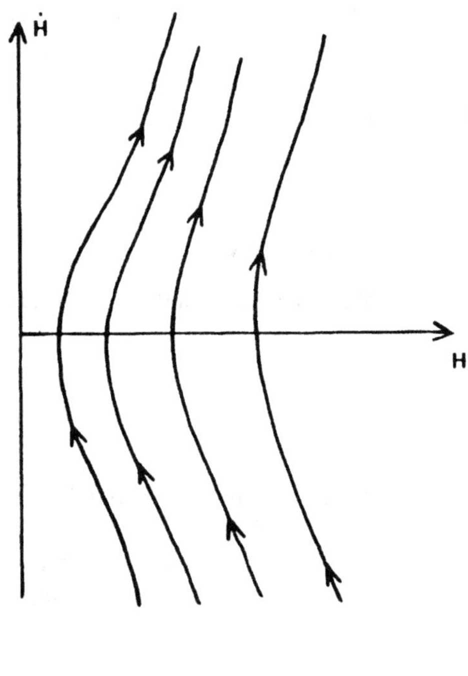

Figure 2. Trajectories without fixed points ($D < 0, \beta > 0, \Lambda > 0, \beta' > 0$).

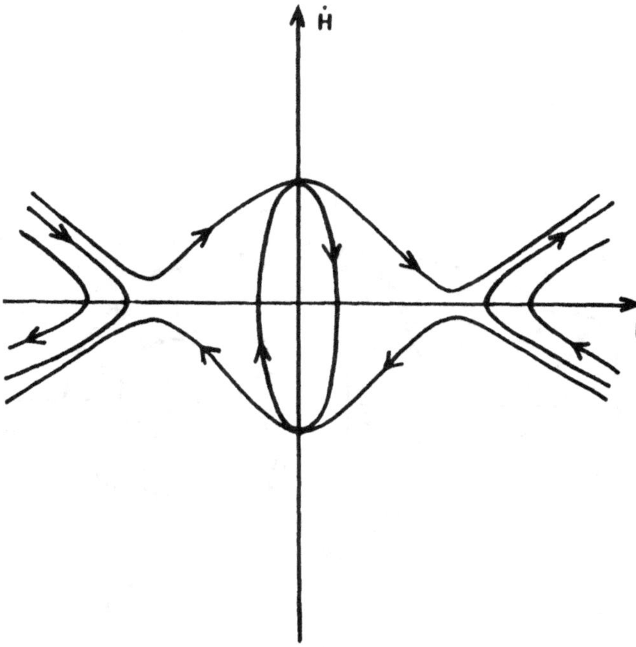

Figure 3. Trajectories with two fixed points ($D > 1, \Lambda > 0, \beta > 0, \beta' < 0$, or $D > 1, \Lambda < 0, \beta > 0, \beta' > 0$).

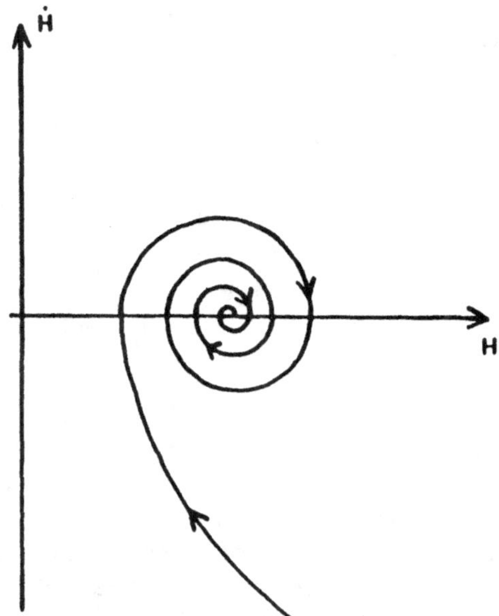

Figure 4. Trajectories with two fixed points ($D > 1$, $\Lambda < 0$, $\beta > 0$, $\beta' < 0$, or $D > 1$, $\Lambda > 0$, $\beta < 0$, $\beta' > 0$).

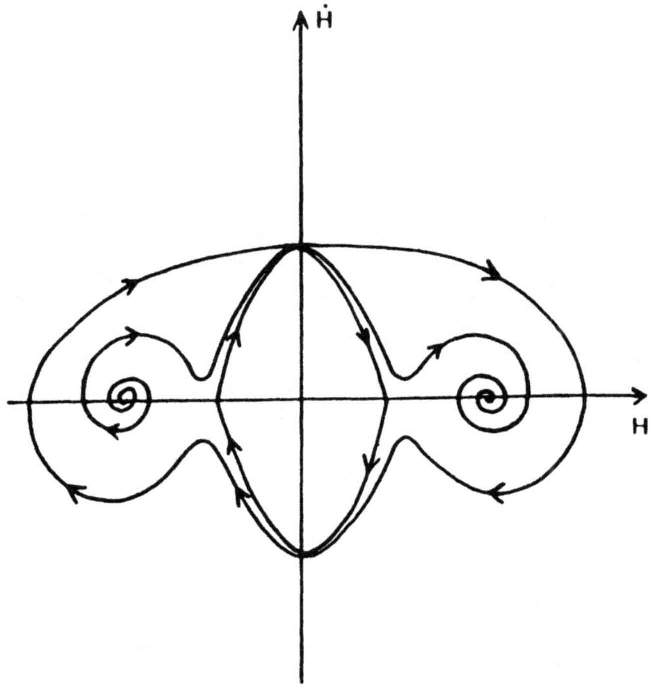

Figure 5. Trajectories with four fixed points ($0 < D < 1$, $\Lambda > 0$, $\beta > 0$, $\beta' > 0$).

expression:

$$\langle T^{\mu}_{\mu}\rangle_{\text{ren}} = \frac{R^2}{34560\pi^2}(N_0 + \tfrac{11}{2}N_{1/2}) + \frac{R}{288\pi^2}(-m_0N_0 + \tfrac{1}{2}N_{1/2}m_{1/2})$$

$$+\frac{R}{48\pi^2}\left\{\frac{m_0^2M_0^2}{2}\cdot N_0\left\{\frac{\psi(3/2+v_0)+\psi(3/2-v_0)}{2} - \ln M_0\right\}\right.$$

$$\left.+\, m_{1/2}^2\cdot N_{1/2}\cdot(1+M_{1/2}^2)[\ln M_{1/2} - |\mathrm{Re}(iM_{1/2})]\right\}$$

$$M_i^2 = 12\frac{m_i^2}{R}, \qquad v_i^2 = \frac{1}{4} - M_i^2 + s_i(s_i - 1)$$

Using this expression we can study the position of the fixed points as a function of the mass. For example, if the mass of the scalar field is increased and we begin (when $m_0 = 0$) with four fixed points (Figs. 5 and 6), two of them move towards infinity and the other two move towards the

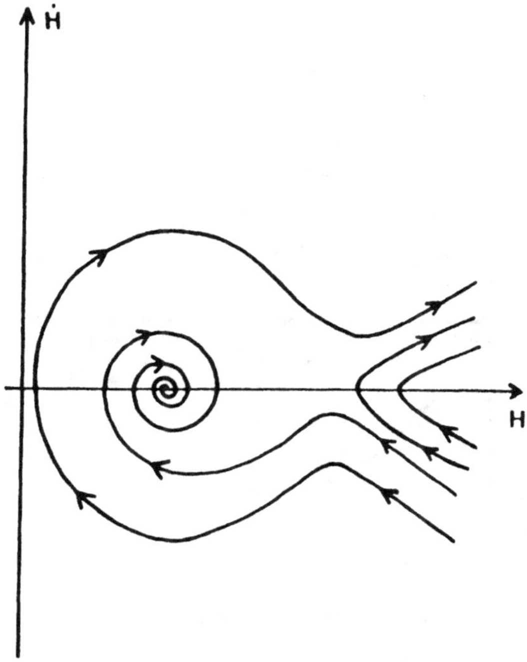

Figure 6. Trajectories with four fixed points ($0 < D < 1$, $\Lambda > 0$, $\beta > 0$, $\beta' < 0$).

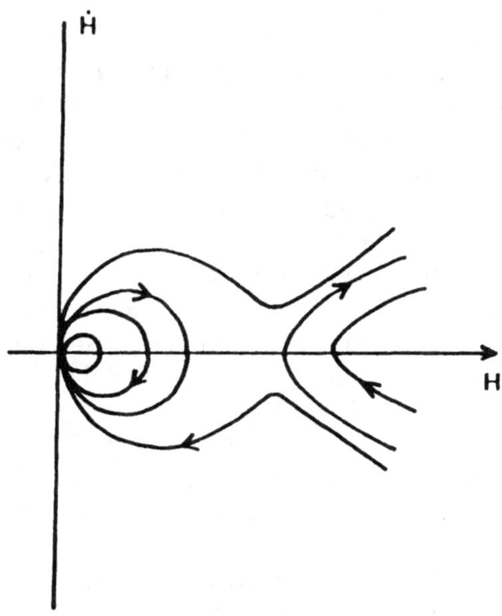

Figure 7. The $\Lambda = 0$ case ($\beta' > 0$).

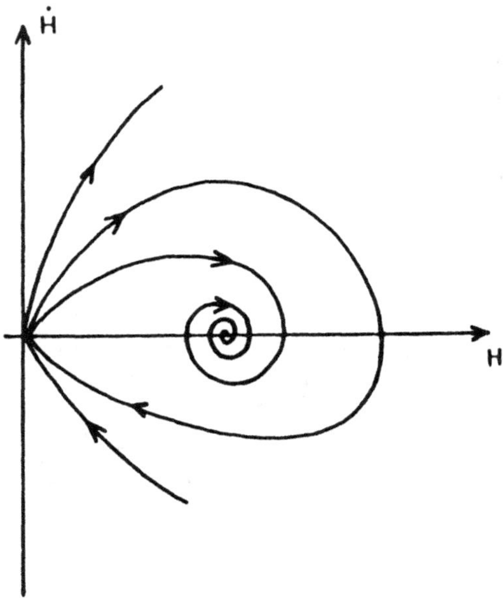

Figure 8. The $\Lambda = 0$ case ($\beta' < 0$).

origin. If we begin with two fixed points (Figs. 3 and 4) both move towards infinity when the mass is increased. Finally, if we begin without fixed points, increasing the mass of the scalar field we can generate four fixed points (we can pass from a situation corresponding to Figs. 1 and 2 to one corresponding to Figs. 5 and 6). We conclude that in this aspect the mass of the field does not introduce any new qualitative behavior. (It can be seen that if we introduce also nonconformally coupled fields we may generate new qualitative behavior because we can pass from a situation with four fixed points when $m_0 = 0$ and $\zeta = \frac{1}{6}$ to a situation with six fixed points when $m_0 \neq 0$ and $\zeta \neq \frac{1}{6}$. In this chapter we restrict ourselves to the case $\zeta = \frac{1}{6}$.)

The modification of the stability of the fixed points (or the shape of the trajectories near these points) is an important subject that is not completely studied. We can mention that in the particular case $\Lambda = 0$ the form of the trajectories near the origin is not modified by the presence of the mass and only depends on the sign of β' (see Figs. 7 and 8) [4] (in Ref. 4 we made the same mistake as Ref. 2 concerning the stability of the Minkowski space; however the result can be reinterpreted in the present context, as we said before).

The complete problem can be treated numerically in the following way: if one of the scalar fields in (2) is a massive one, the renormalized expectation value of the energy density can be written as (see Ref. 5 and references therein)

$$\langle 0|T_{00}|0\rangle_{\text{ren}} = T^{I}_{00} + \int \frac{d^3k}{(2\pi a)^3}\,\omega_k|\beta_k(t)|^2 - \frac{m^2}{288\pi^2}\left(R_{00} - \tfrac{1}{2}Rg_{00}\right) \tag{4}$$

where

$$\dot{\beta}_k(t) = \frac{m^2 H}{2\omega^2}\,\alpha_k(t)\exp\left(2i\int^t \omega_k\,dt'\right) \tag{5a}$$

$$\dot{\alpha}_k(t) = \frac{m^2 H}{2\omega^2}\,\beta_k(t)\exp\left(-2i\int^t \omega_k\,dt'\right) \tag{5b}$$

$$|\alpha_k(t)|^2 - |\beta_k(t)|^2 = 1$$

$$\omega_k = \left(\frac{k^2}{a^2} + m^2\right)^{1/2}$$

(the other components can be deduced from this expression and the conservation rules). The back-reaction problem in this case can be completely studied solving simultaneously equation (1) with (4) as the source and equations (5). These equations (5a) and (5b) are equivalent to the field equation for the scalar field (the Klein–Gordon one). One question that immediately arises is the meaning of the initial conditions for $\alpha_k(t)$ and $\beta_k(t)$. A particular initial condition is related to the selection of a basis for the set of solutions of the Klein–Gordon equation, and this is obviously related to a particular vacuum definition. A useful vacuum definition is the one related to the conditions $\alpha_k(t_0) = 1$ and $\beta_k(t_0) = 0$: the vacuum that is defined is the one that minimizes the Hamiltonian at $t = t_0$. However, we can easily relate two expectation values of the energy-momentum tensor evaluated in two different vacua $|0_1\rangle$ and $|0_2\rangle$:

$$\langle 0_1|T_{00}|0_1\rangle_{\text{ren}} = \langle 0_2|T_{00}|0_2\rangle_{\text{ren}} + \int \frac{d^3k}{(2\pi a)^3}\,\omega_k|\tilde{\beta}_k|^2 \tag{6}$$

where $\tilde{\beta}_k$ is one of the Bogoliubov coefficient that relates the basis associated to each vacuum state.

It is obvious that in the massless case the second term on the right-hand side of (6) is nothing but the energy density of a fluid of classical radiation ($T_{00} \sim a^{-4}$). In the massive case this term is also the energy density of a fluid of classical matter with an energy spectrum given by $|\tilde{\beta}_k|^2$.

Then the general problem of the back reaction can be treated by putting in the right-hand side of (1) the tensor that is obtained as the sum of (I) the tensor given in (2), which is the contribution of the conformally invariant fields; (II) one term of classical radiation which takes into account the possibility that the quantum state would be different from the conformal vacuum (the energy spectrum is related to the coefficient of the correspondent Bogoliubov transformation); (III) the terms that appear in (4) that take into account the particle creation for the scalar field (an analogous treatment could be done for the spin $\frac{1}{2}$ fields); (IV) one term of classical matter with a given energy spectrum which takes into account the possibility that the quantum state of the massive scalar field would be a vacuum state different from the one defined by the initial condition over $\alpha_k(t)$ and $\beta_k(t)$.

REFERENCES

1. T. Azuma and S. Wada, General spatially flat cosmological solution to the Einstein equation with the cosmological constant and the anomaly, University of Tokio Komaba Report (1984).
2. A. Vilenkin, *Phys. Rev. D 32*, 2511 (1985); A. A. Starobinski, in: *Quantum gravity* (M. Markov and P. West, eds.), Plenum Press, New York, 1984, p. 103.
3. M. Castagnino, D. Harari, and J. P. Paz, *Class. Quantum Grav. 3*, 569 (1986).
4. M. Castagnino and J. P. Paz, *Phys. Lett. 164B*, 274 (1986).
5. M. Castagnino, D. Harari, and C. Nuñez, Vacuum *J. Math. Phys. 28*, 184 (1987).

Chapter 8

Gauge Theory of Conformal Group

Mario Castagnino and Claudia Yastremiz*

In this chapter we try to induce, in the most natural possible way, space-time structures from internal ones. This is done through the mathematical notion of soldering, and as an application example, we show how we constructed gravity starting with the gauge fields of the conformal group.

So, in the first place we show more or less trivial mathematical examples of the soldering, to see what the notion means. Then we state the formal definition and a couple of useful theorems, and in the end we sketch the deduction of gravity from the conformal gauge fields.

Let G be one of the affine groups [the general affine group $GA(n, R)$, the Poincaré group $AO(\nu+, \nu-)$, the Weyl group $AS(\nu+, \nu-)$, etc.] and G' its isotropy group in some point $O \in R$. Suppose now we have an n-dimensional differentiable manifold M. Then the homogeneous space $G/G' \simeq R^n$ is isomorphic to $T_x(M)$ for each $x \in M$ [1], so if one constructs the bundle $E(M)$ with standard fibre $F \simeq G/G'$, one can identify $T(M)$ with $T_F(E)$. Moreover, if one has an affine connection, its 1-form ω can be decomposed uniquely as $\omega = \omega^R + \omega_0$, where ω^R goes to the Lie algebra

* Fellow of the Consejo Nacional de Investigaciones Científicas y Técnicas.

MARIO CASTAGNINO • Instituto de Astronomía y Física del Espacio, 1428 Buenos Aires, Argentina; and Instituto de Física de Rosario, 2000 Rosario, Argentina. CLAUDIA YASTREMIZ • Instituto de Astronomía y Física del Espacio, 1428 Buenos Aires, Argentina.

of G' and is the 1-form of a linear connection and ω_0' goes to R. Once a gauge is chosen, it is an $n \times n$ matrix, and so it is suitable as a matrix for the isomorphism between $T_F(M)$ and R^n (if it is invertible). We called this a trivial example because one is usually interested only in the linear connection, and the standard election is to take ω_0 as the canonical 1-form, that is, the identity matrix, so it is barely noticed. Other choices look like a basis change in the tangent space, and they go from holonomic (tangent) to antiholonomic (G/G') indices; ω_0 components are then called e_μ^m and that is nothing more than the vierbein (no orthonormality condition yet!).

However, in the general case one must be careful, because it could be that $G/G' \neq R$ [but its tangent space can still be n dimensional and so be isomorphic to $T_z(M)$ at each point]. So we proceed (with the affine examples in mind) to state the formal definition.

Let $P(M, G)$ be a principal bundle, g the Lie algebra of G, and $E(M, F, G, P) = E(M)$ an associated bundle to P with standard fibre F. We say that $E(M)$ is presoldered to M [2] if the following conditions hold:

1. G acts transitively on F (given two elements of F there is an element of G that carries one to another).
2. $\dim F = \dim M$.
3. $E(M)$ admits a cross section: $E \to M$ (supposed given).

Condition 1 implies that F is a homogeneous space, isomorphic to the quotient G/G' (with G' the isotropy group of G in $O \in F$), and therefore that $g \simeq g' \in T_0(F)$ (semidirect sum). Condition 3 is equivalent to giving a reduction from G to G' [that is, to defining a reduced subbundle $P'(M, G')$, which is a principal bundle of structural group G' and differential structure compatible with that of P].

We say that $E(M)$ is solderable to M if it is presoldered and moreover it verifies the following condition:

4. The bundles $T(M)$ and $T_F(E)$ are equivalent, in the sense that there exists a diffeomorphism Σ: $T(M) \to T_F(E)$ such that restricted to each fiber is an isomorphism of vectorial spaces. Here $T_F(E)$ is the subbundle of $T(E)$ given by the vertical vectors (that is, tangent to the fiber) at the points of $\sigma(M)$. That is,

$$T_F(E) = \bigcup_{x \in M} \{T_{(x)}(F_x)\}$$

Each equivalence Σ is called a soldering (it is called "soudure" in some mathematical literature) from $E(M)$ to M. If condition 4 is true, σ is a soldering cross section (it amounts to selecting an origin in each fiber), and $E(M)$ is soldered to M if it is solderable and a soldering has been chosen.

The following theorems, given without proof, show the existence of soldering constructively:

Theorem 1. If $E(M)$ is presoldered to M, then $E(M)$ is solderable to M iff there exists a 1-form

$$\omega_0: T(P') \to T_0(F)$$

[where $P' = P'(M, G)$ is given by the cross section σ] such that

a. ω_0 is a tensorial 1-form of type (iso, $T(F)$), with iso: $G' \to GL(T_0(F))$ the linear isotropic representation of G', $\text{iso}(a) = (L_a)_*$:

- $\forall u \in G'$;
- $\omega_0[(R_a)_* X_u] = \text{iso}(a^{-1})\omega_0(X_u) \qquad \forall X_u \in T_u(P')$;
- X_u vertical implies $\omega_0(X_u) = 0$;

b. $\forall X_u T(P')$, $\omega_0(X_u) = 0$ implies $X_u = 0$.

Moreover, there exists a one-to-one correspondence between solderings from $E(M)$ to M and 1-forms ω_0, given by

$$\omega_0(X_u) = u_*^{-1}(\Sigma(X_x)) \quad \text{and} \quad \Sigma(X_x) = u_*(\omega_0(X_u))$$

where $X_x = \pi_*(X_u)$, $\pi: P' \to M$ is the projection that defines the bundle, and $u_*: T_0(F) \to T_{\sigma(x)}(F_x)$ is the isomorphism induced by the diffeomorphism $u: F \to F_x$ (that assigns to each tangent vector of the standard fiber the tangent vector to the fiber over x that has the same components in the referential u, so its matrix is the identity). The 1-forms that satisfy (a) and (b) are the soldering 1-forms.

Definition. A Cartan principle bundle of type F is a bundle $P(M, G)$ such that there exists an associated bundle $E(M, F, G, P)$ solderable to M and M is paracompact. If $P(M, G)$ is a Cartan principal bundle of type F, a Cartan connection in P is a connection such that its 1-form $\omega: T(P) \to g$ verifies $X_u \in T_u(P')$ and $\omega(X_u) = 0$ imply $X_u = 0$. That is, we demand that there must not be nonzero horizontal vectors in $T(P')$ (all the infinitesimal transformations move around in the fiber and do not induce movements on the base manifold).

Theorem 2. Let $P(M, G)$ be a Cartan principal bundle of type F and let $\omega: T(P) \to g$ be the Cartan connection 1-form. Then, if $\gamma_0: g \to T_0(F)$ is the projection of g onto its subspace $T_0(F)$, $\omega_0 = \gamma_0 \circ \omega$ is a soldering 1-form from $E(M)$ to M.

Clearly, then, it suffices to give a connection in the principal bundle, and to find in the Lie algebra of the group a semidirect summand with the same dimension as the base manifold to give the soldering, that is, an isomorphism between that semidirect summand and the tangent space at each point of M. That gives the transformation between the "internal" or "anholonomic" indices and the "external" or "holonomic" ones. Because of the expression of Σ in terms of ω_0 (remember that the u_* matrix is the identity) and of ω_0 in terms of ω, the components of the isomorphism are exactly the components of ω in the direction of $T_0(F)$ (that is, the vierbein). The condition of being a Cartan connection is that the vierbein is nonsingular seen as a matrix.

So let us now start with the example. We have chosen the conformal group because it is the smallest semisimple group that contains the Poincaré group as a subgroup, and moreover, it includes the de Sitter groups and the scale transformations, which are known to play an important role in physical theories, at least at high energies.

So we have the conformal group $C(1,3)$ $[SO(4,2)\ SU(2,2)]$ with the usual commutation relations between the generators P_m (translations), K_m (conformal boosts), M_{mn} (Lorentz generators), and D (dilatations):

$$
\begin{aligned}
[P_m, P_n] &= [K_m, K_n] = [M_{mn}, D] = 0 \\
[P_m, M_{rs}] &= \eta_{ms} P_r - \eta_{mr} P_s \\
[K_m, M_{rs}] &= \eta_{ms} K_r - \eta_{mr} K_s \\
[K_m, P_n] &= -2(\eta_{mn} D + M_{mn}) \\
[P_m, D] &= P_m \\
[K_m, D] &= -K_m \\
[M_{mn}, M_{rs}] &= \eta_{rm} M_{ns} - \eta_{rn} M_{ms} + \eta_{sn} M_{mr} - \eta_{sm} M_{nr}
\end{aligned}
$$

Given the structure constants, one then can define the Killing metric of the Lie algebra as $\gamma_{AB} = C^{C}_{AD} C^{D}_{CB}$, which is invertible owing to the semisimplicity of the group. The calculation gives

$$
\begin{aligned}
\gamma_{m\hat{n}} &= -8\eta_{mn} \\
\gamma_{[mn][rs]} &= +8(\eta_{rp}\eta_{sq} - \eta_{rq}\eta_{sp}) \\
\gamma_{..} &= -8
\end{aligned}
$$

(the caret is the index for the K, the dot for D, and the others are null).

Let us suppose we have a four-dimensional manifold M that has only its differential structure, and construct a principal bundle $P(M, C(1,3))$ with this "internal" conformal group as structure group. Suppose then we have a connection in this bundle and we choose a gauge, that is, a section σ. Then the gauge fields are the components of $\omega_\sigma = \sigma^*\omega$ (ω is the connection 1-form): $\omega_\sigma = \omega_\mu x^\mu$

$$\omega_\mu = e_\mu^m P_m + f_\mu^m K_m + \tfrac{1}{2}\omega_\mu^{mn} M_{mn} + b_\mu D$$

Then the field strength is the curvature of the connection,

$$R_{\mu\nu}^A = h_{\mu,\nu}^A - h_{\nu,\mu}^A + C_{BC}^A h_\mu^B h_\nu^c$$

We want to induce from this internal structure only the external metric and the linear connection on M, so we must realize the soldering. As we have defined a connection (which will be determined dynamically), the only things we need are the generators of the space $T_0(F)$. The only possibility that preserves local Lorentz invariance is a linear combination of P and K, $P'_m = \alpha P_m + \beta K_m$. We take then as a new basis of the Lie algebra the set P'_m, K_m, M_{mn}, and D, so $\alpha \neq 0$. Following the formulas given in the definitions and using the fact that an isomorphism between the vector spaces induces an isomorphism between any of the corresponding tensor spaces associated with them, there is an induced metric on M, $g_{\mu\nu} = -16\alpha\beta\eta_{mn}e_\mu^m e_\nu^n$, where $-16\alpha\beta\eta_{mn}$ is the Killing metric restricted to $T_0(F)$. In order for **g** to be nonsingular, it must be $\beta \neq 0$ too, and that shows that the relevant group at low energies (where scale invariance and conformal boosts do not usually appear) is de Sitter (or anti-de Sitter) and not Poincaré. That was assumed by McDowell and Mansouri (without explaining why they do not use the more natural election of the Poincaré group) in their paper of 1977 [3]. In this case, β (or better β/α) looks like a mixing angle between P and K.

We can define a covariant derivative too, by taking the vector to the inner space through the soldering, taking the derivative there, and going back to the tangent space. With this definition, asking for consistency and for the Leibniz rule to apply, results in an equation for the vierbein that is exactly the "tetrad postulate":

$$D_\nu X^\mu = (e^{-1})_m^\mu D_\nu(e_\lambda^m x^\lambda) \Rightarrow D_\nu e_\lambda^m = 0 \Rightarrow \partial_\nu e_\mu^m - \Gamma_{\mu\nu}^\lambda e_\lambda^m + \omega_\gamma^{rm}\eta_{sr}e_\mu^s$$
$$+ b_\nu e_\mu^m \Rightarrow \Gamma_{\mu\nu}^\lambda = (e^{-1})_r^\lambda (e_\mu^m \eta_{ms}\omega_\nu^{rs} + e_{\mu,\nu}^r) + \delta_\mu^\lambda b_\nu$$

Then the curvatures and the torsion for this connection can be expressed in terms of the $T_0(F)$ component of the conformal curvature and the part of the Lorentz (mn) components corresponding only to the "spin connection" ω_μ^{mn} as follows:

$$S_{\mu\nu}^\lambda = (e^{-1})_m^\lambda R_{\mu\nu}^m$$

$$R_{\beta\gamma\delta}^\alpha = \eta_{rp}(e^{-1})_f^\alpha e_\beta^r \, {}^E R_{\gamma\delta}^{pq}, \qquad R = \frac{-1}{16\alpha\beta}(e^{-1})_p^\nu (e^{-1})_q^\sigma \, {}^E R_{\nu\sigma}^{pq}$$

To write the Lagrangian, we must not use the space-time metric, because that would make the gauge fields enter in a nonminimal way, so we cannot write directly the Einstein Lagrangian nor the Yang–Mills one. Therefore we can only write something like

$$L = Q_{AB} R^A \wedge R^B$$

but the obvious choice for Q_{AB}, namely, the Killing metric, would give us a topologic invariant, which is useless as a Lagrangian. So we will do here the only thing we consider completely ad hoc, namely, replacing $8(\eta_{rp}\eta_{sq} - \eta_{rq}\eta_{sp})$ by ε_{rspq} [that is what McDowell and Mansouri did for $Sp(4)$ [3] and is the choice of Pagels for $O(5)$] [4, 5]. We will assume, moreover, that the b and f' fields become very heavy because at low energies the scale and conformal boost's invariance is broken (we do not assume any special mechanism). With this assumption, the Lagrangian can be written

$$L = \varepsilon^{\mu\nu\lambda\sigma} g_{\rho\delta} S^{\rho}_{\mu\nu} S^{\delta}_{\lambda\sigma} + 2\sqrt{-g}\, R + 6\sqrt{-g} + \varepsilon^{\mu\nu\lambda\sigma}_{rspq}\ {}^{E}R^{rs}_{\mu\nu} R^{pq}_{\lambda\sigma} + \mathcal{L}'(b, f)$$

where the fourth term is the topologic invariant on M and can be dropped; and calculating the effective action by integrating over the heavy fields [6] can only contribute by changing the constants in front of the first three terms, so the effective Lagrangian is

$$L_{\text{eff}} = C_1 \varepsilon^{\mu\nu\lambda\sigma} g_{\rho\delta} S^{\rho}_{\mu\nu} S^{\delta}_{\lambda\sigma} + c_2\sqrt{-g}\, R + c_3\sqrt{-g}$$

REFERENCES

1. S. Kobayashi and K. Nomizu, *Foundations of Differential Geometry*, Interscience, New York, 1963; M. Spivak, *A Comprehensive Introduction to Differential Geometry*, Publish or Perish, New York, 1965; A. Trautman, in: *General Relativity and Gravitation* (A. Held, ed.), Chap. 9, Vol. I, p. 287, Plenum Press, New York, 1980.
2. S. Kobayashi, *Can. J. Math. 8*, 145 (1956).
3. S. W. McDowell and F. Mansouri, *Phys. Rev. Lett. 38*, 739 (1977).
4. H. R. Pagels, *Phys. Rev. D 27*, 2299 (1983).
5. H. R. Pagels, *Phys. Rev. D 29*, 1690 (1984).
6. S. L. Adler, *Rev. Mod. Phys. 54*, 729 (1982).

Chapter 9

Three Lectures on String Theory

David J. Gross

1. A BROAD REVIEW OF STRING THEORY

Three lectures on string theory are clearly too few, because string theory is already an enormous field involving new concepts and advanced mathematical techniques. In three lectures it is only possible to skim the surface of this rapidly developing theory, pointing out the main features and explaining, at a pedestrian level, some of the important ideas. In the first lecture, Section 1, I give a "pep-talk," a broad review of string theory: why it is interesting, where it stands, and where it is going. In the subsequent lectures, Sections 2 and 3, I discuss a few specialized topics in more detail.

High-energy physics is, at present, in an unusual state. It has been clear for some time that we have succeeded in achieving many of the original goals of particle physics. We have constructed theories of the strong, weak, and electromagnetic interactions. The "standard model" is remarkably successful and seems to be an accurate and complete description of all of physics, at least at energies below 1 TeV. Indeed, there are at the moment no significant experiments that cannot be explained by the color gauge theory of the strong interactions (QCD) and the electroweak gauge theory.

DAVID J. GROSS • Joseph Henry Laboratories, Princeton University, Princeton, New Jersey 08544.

New experiments continue to confirm the predictions of these theories and no new phenomenon has appeared.

This success has not left us sanguine. Our present theories contain too many arbitrary parameters and unexplained patterns to be complete. They do not satisfactorily explain the dynamics of chiral symmetry breaking or of CP symmetry breaking. The strong and electroweak interactions cry out for unification. Finally, we must ultimately face up to the task of including quantum gravity within the theory. However, we theorists are in the unfortunate situation of having to address these questions without the aid of experimental clues. Furthermore, extrapolation of present theory and early attempts at unification suggest that the natural scale of unification is at least 10^{16} Gev, tantalizingly close to the Planck mass scale of 10^{19} Gev. It seems very likely that the next major advance in unification will include gravity. I do not mean to suggest that new physics will not appear in the range of TeV energies. Almost all attempts at unification do, in fact, predict a multitude of new particles and effects that could show up in the TeV domain (Higgs particles, Supersymmetric partners, etc.), whose discovery and exploration is of the utmost importance. But the truly new threshold might lie in the totally inaccessible Planckian domain.

What strategy are theorists to adopt under these circumstances? One (defeatist) strategy is to do nothing until experiment provides more clues or paradoxes. The other (dangerous) is to push forward and attempt to make a bold leap into the unknown, searching for new symmetries and unifying principles. We are, however, not leaping totally into the dark. Any attempt at further unification is highly constrained by the requirement that it be consistent with the very large amount of understanding that we have of "low-energy" (≤ 1 TeV) physics. This provides powerful phenomenological constraints. Furthermore, we do have one window into Planck mass physics, namely, we know much about gravity, a force whose typical energy scale is the Planck mass. The necessity of including gravity in our theoretical framework provides powerful theoretical constraints.

What are the guidelines in the search for higher unification? The pioneering attempts were modest in scope. Based on the striking similarities of the strong and electroweak gauge interactions and the way that the quark and lepton multiplets seemed to be related, one tried to enlarge the gauge group of nature and construct at a high-energy scale a single grand unified gauge interaction. These attempts have largely failed. First, the simple and striking predictions that they made, in particular a proton lifetime of 10^{32}–10^{33} years, have not been verified. Second, although much good physics came out of these attempts (a scenario for proton and the cosmological production of baryons, the study of massive neutrinos, the idea of inflation, etc.), it also became clear that their scope was too limited. Realistic GUT models contained even more arbitrary parameters and required more fine tuning than the standard theory and the mechanism for spontaneous symmetry breaking had to be put in by hand.

GUT theories also brought to the forefront a new and serious issue—the *hierarchy* problem, namely, why is the ratio of the W mass to the GUT mass scale (M_G) so small? This is not a new problem; it had worried Dirac many years ago (why is the proton mass so much smaller than the Planck mass?), but in the context of GUT theories it is more worrisome. The point is that in such theories the only natural scale is the unification scale, so that all massive particles should have a mass of M_G. The only known way to avoid this is to imagine a symmetry principle that forbids a mass from developing. If this symmetry is broken at a scale much smaller than M_G then much smaller masses can naturally arise. In the case of fermionic matter (quarks and leptons) this scenario is easy to arrange. Chiral symmetries protect spin one-half fermions from developing masses (a mass term, $\bar{\psi}\psi$, is not invariant under chiral transformations). Such symmetries are indeed a property of the observed world (the V − A electroweak interactions are not chirally symmetric). If we can protect this symmetry from breaking until the scale of M_W then we can understand the smallness of both the W mass and the quark and lepton masses (in this game one assumes that the mass ratios of quark and lepton masses to M_W can be explained, by powers of small couplings). The problem is that the breaking of chiral gauge symmetries is generated by the vacuum expectation values of Higgs particles. But these are scalar particles whose vanishing mass is not protected by chiral symmetries. So what ensures that these do not develop expectation values of order M_G, leading to $M_W \approx M_G$?

The most elegant approach to solving this problem is by means of supersymmetry. Supersymmetry relates fermions to bosons, quarks to Higgs particles. It puts the scalar particles in the same multiplet as the fermions and protects masses from developing as a consequence of supersymmetry. One then has to imagine that supersymmetry remains unbroken until low energy (of order 1 TeV). This is not too hard to imagine since, in fact it is quite difficult to break supersymmetry at all. In many theories, once introduced, it can only be broken nonperturbatively. Nonperturbative effects are energy dependent (as in QCD, where confinement "turns on" at large distance), and are governed by dimensional couplings that vary logarithmically with energy. Thus one can plausibly imagine that M_W/M_G arises as $\exp[-1/g^2]$, where g is some small gauge coupling at the unification scale.

This possible solution of the hierarchy problem is the strongest physical reason for trying to build supersymmetry into a unified theory. There are, of course, other reasons. If there is one lesson to be learned from the success of the last decades, it is that the secret of nature is symmetry. If one is to make new progress in unification one must discover new symmetries of nature. This is not trivial. In order to discover truly new symmetries one must discover new degrees of freedom of nature, on which the new symmetries can act, as well as dynamical mechanisms that hide or break these symmetries. After all, if the symmetries were manifest they would be known already. Supersymmetry is the most attractive of the new symmetries that one can imagine. One can think of this symmetry as a consequence of the enlarging of configuration space from ordinary space to a "superspace" which contains fermionic coordinates. Supersymmetry is then a consequence of a natural generalization of ordinary space-time symmetries to superspace. It also has a natural local version—supergravity. This marvelous new symmetry is very powerful. I find its most attractive feature to be the possibility of it yielding an "explanation" of the existence and nature of matter. In the standard model gauge particles (gluons, W's and Z's) are automatic consequences of local gauge symmetry. Quarks and leptons must be put in by hand. This is ugly and asymmetrical and leads to much arbitrariness. In supergauge theories, one might imagine that the matter is in supersymmetric multiplets together with gauge mesons, and thus exists as a consequence of gauge symmetry. One could therefore hope that supersymmetry would predict the nature of fermionic matter, and also the spectrum of scalar (Higgs) particles, in a way that could solve the hierarchy problem.

This would still leave us with the question as to why nature chooses the particular ($SU_3 \times SU_2 \times U_1$) gauge group that we observe. A possible answer to this question might be provided by the revival of an old idea that was put forward by Kaluza in 1921! Kaluza, shortly after Einstein developed his theory of gravity, considered the possibility that there might exist more than three spatial dimensions. He noted that under certain circumstances the extra dimensions could lead naturally, in the framework of a purely gravitational theory, to Maxwell's theory of electromagnetism and the emergence of a conserved and quantized electric charge. This possibility has been generalized to non-Abelian gauge groups and explored in great detail in recent years. It is now clear that the question of how many spatial dimensions there are is an experimental one, and that there is no conflict with present observation if the extra dimensions are compactified into a small (less than 10^{-15} cm) manifold. Kaluza's original idea was a five-dimensional world with pure gravitational interactions. One can find solutions of Einsteins equations with the topology of four-dimensional Minkowski space-times a circle of small radius R. At low energies compared to $1/R$ one would not notice the extra dimension, which requires energies of $1/R$ to excite. However, there would exist a new conserved quantity, P_5, the component of the energy-momentum in the fifth direction. Now, full five-dimensional Poincaré invariance is broken by our asymmetric vacuum; however, one still has a symmetry of translations about the hidden circle. In fact, this is a local gauge symmetry. Consequently P_5 is conserved (and, since the circle is compact, quantized) and furthermore it is coupled to a massless gauge boson (which corresponds to one of the polarizations of the graviton along the hidden dimension). We can, following Kaluza, identify P_5 with a multiple of the electric charge and the gauge boson with the photon. We thus have *derived* electromagnetism and the quantization of electric charge from five-dimensional gravity.

This idea is easily generalized to theories with more than five dimensions, in which case non-Abelian gauge groups can emerge from compactified vacua. If we combine this idea with supersymmetry one might dream of a higher-dimensional supergravity theory which produces, upon dynamical compactification, the low-energy gauge groups of nature while at the same time determining the matter content—in other words a predictive theory of everything. These ideas have been pursued at length over the last decade. However, it appears that in the context of ordinary quantum field theory, which heretofore has served us so well, they cannot be made to work. This is for many reasons, but three stand out. It is useful to recall these, if only to compare with the early successes of string theory.

The first problem is that the quantum field theory of gravity arrived at by straightforward quantization of Einstein's theory is sick (nonrenormalizable) at high energies. This sickness, which is not cured by supersymmetry or other modifications of the theory, means that the physics must be modified at energies of order the Planck mass. It is clearly dangerous to try to construct a unified theory on such shaky foundations. One might, however, argue that the high-energy modifications will not affect low-energy physics and proceed to construct realistic models. One found, however, at least in the context of pure Kaluza-Klein supergravity, that one could not construct realistic models, for two reasons. First, it appeared to be impossible to understand the emergence of chiral fermions. The chiral nature of quarks and leptons, the fact that the weak interactions are V − A and there are no V + A weak interactions, is one of the most striking features of nature. It is a very important feature since it is what keeps these fermions light; otherwise they would have gotten an enormous mass and we would never have seen them. It is not easy to get chiral fermions from a geometrical theory of matter, since there is nothing in ordinary geometry that picks out one handedness. Within the framework of the most ambitious Kaluza-Klein supergravity theories it appears impossible to derive chiral matter multiplets. Second, the background energy of the vacuum, otherwise known as the cosmological constant Λ, always came out to be of order M_P^4. This is in violent contradiction to observation, which indicates that $\Lambda \leq 10^{-120}\, M_P^4$. For these and other reasons no one ever came up with realistic field theoretic models based on these ideas.

This brings us to string theories, in particular the heterotic string, which appears to suffer from none of these problems. It yields a finite and consistent theory of gravity, in which chiral fermions and realistic gauge interactions can naturally emerge and in which, for most of the existing solutions, the cosmological constant vanishes. It incorporates supersymmetry, higher dimensions, and much more!

String theories provide a way of realizing the potential of supersymmetry, Kaluza–Klein unification, and much more. They represent a radical departure from ordinary quantum field theory, but in the direction of increased symmetry and structure. They are based on an enormous increase in the number of degrees of freedom, since in addition to fermionic coordinates and extra dimensions, the basic entities are extended one-dimensional objects instead of points. Instead of fields, or wave functions, that are functions of coordinates x^μ we have fields that are functionals of loops $\Psi[x^\mu(\sigma)]$. Whereas a quantum field describes naturally one particle at a time, a string field will describe an infinite number of particles, with a degeneracy that increases exponentially with mass. You might think that such an increase in the size of configuration space is wasteful and unnecessary. After all, we only observe a finite number of particles in nature; why do we need so many degrees of freedom? The answer is that corresponding to this increase there is an increase in the symmetry group of nature, in a way that we are only beginning to comprehend. At the very least, this extended symmetry group contains the largest group of symmetries that can be contemplated within the framework of point field theories—those of ten-dimensional supergravity and super Yang–Mills theory.

The origin of these symmetries can be traced back to the geometrical invariance of the dynamics of propagating strings. Traditionally string theories are constructed by the first quantization of a classical relativistic one-dimensional object—whose motion is determined by requiring that the invariant area of the world sheet it sweeps out in space-time is extremized. In this picture the dynamical degrees of freedom of the string are the coordinates, $x_\mu(\sigma, \tau)$ (plus fermionic coordinates in the superstring), which describe its position in space time. The symmetries of the resulting theory are all consequences of the reparametrization invariance of the σ, τ parameters which label the world sheet. As a consequence of these symmetries one finds that the free closed string contains a massless spin-two meson, which can be identified as the graviton, whereas the open string, which has ends to which charges can be attached, yields massless vector mesons, which can be identified as Yang–Mills gauge bosons.

Let us construct the theory of a free relativistic open string. This is an extended object in space. To describe it in an invariant manner we note that as the string moves it sweeps out a two-dimensional strip in Minkowski space, which can be parametrized by $0 \leq \sigma \leq \pi$ and τ, so that as τ runs from τ_1 to τ_2, $x^\mu(\sigma, \tau)$, which describes the string position in Minkowski space, runs from $x_1^\mu(\sigma)$ to $x_2^\mu(\sigma)$. The natural generalization of the length of the world line is the area of the world sheet swept out by the moving string. Therefore we take the action to be the "Nambu action"

$$S = \text{invariant area}$$

$$= -T \int_{\tau_1}^{\tau_2} d\tau \int_0^{\pi} d\sigma \left[\left(\frac{dx}{d\tau} \cdot \frac{dx}{d\sigma} \right)^2 - \left(\frac{dx}{d\tau} \right)^2 \cdot \left(\frac{dx}{d\sigma} \right)^2 \right]^{1/2} \tag{1}$$

and determine the (classical) motion of the string so as to extremize this area. T is the "tension" of the string and will turn out to be of order M^2_{Planck}. [One must also specify boundary conditions at the ends of the string, the appropriate ones being $dx^{\mu}(\sigma)/d\sigma = 0$, for $\sigma = 0, \pi$.]

This action is invariant under a large group, that of reparametrizations of both world sheet labels, σ and τ. Correspondingly only $D-2$ of the string coordinates are physical (D is the dimension of the space-time in which the string moves). One coordinate can be identified with the time coordinate and the longitudinal degree of freedom is just a relabeling of the string. This leaves us with $D-2$ physical degrees of freedom, which describe the oscillations of the string transverse to itself. At the price of breaking manifest covariance it is sometimes useful to eliminate the unphysical degrees of freedom. To do this we identify one of the string coordinates with the timelike label of the string world sheet, τ. We shall choose not the obvious time coordinate $x^0(\sigma, \tau)$, but rather the "light cone time," $x^+ \equiv 2^{-1/2}(x^0 + x^{D-1})$, as measured by an observer moving at the velocity of light along the $D-1$ axis. The advantage of this choice can be seen from the particle case, where, if we choose $x^+ = \tau$, then the conjugate momentum $p^- \equiv 2^{-1/2}(p^0 - p^{D-1})$ which generates light cone time translations (the light cone gauge Hamiltonian) is given by $p^- = (1/2p^+)[(p^i)^2 + m^2]$, thereby eliminating the nasty square roots that occurred before.

Therefore we use the reparametrization freedom to choose

$$x^+(\sigma, \tau) = x^+ + P^+\tau, \qquad p^+(\sigma, \tau) = P^+ \tag{2}$$

One then finds that $x^-(\sigma, \tau)$ and the Hamiltonian, P^-, can be expressed in terms of the transverse degrees of freedom x^i, $i = 1, 2, \ldots, D-2$

$$P^- = \frac{1}{2P^+} \int_0^{\pi} d\sigma \left[\left(\frac{dx^i}{d\tau} \right)^2 + \left(\frac{dx^i}{d\sigma} \right)^2 \right] \tag{3}$$

This is a rather simple Hamiltonian, which can easily be quantized. After all it is just a bunch of harmonic oscillators. The equation of motion for x^i is simply the two-dimensional wave equation, $d^2x^i/d\tau^2 - d^2x^i/d\sigma^2 = 0$, and the general solution can be expanded as

$$x^i(\sigma, \tau) = x^i + p^i\tau + \sum_{n=1}^{\infty} (a^i_n e^{-in\tau} + a^{i\dagger}_n e^{in\tau}) \frac{\cos n\sigma}{\sqrt{n}} \tag{4}$$

Upon quantization $a_n^\dagger$ and a_n are creation and annihilation operators for the normal modes of vibration of the string, which obey the canonical commutation relations, $[a_n^i, a_m^{j\dagger}] = \delta_{i,j}\delta_{n,m}$, of harmonic oscillators. Thus the Hilbert space of the free relativistic string is given by states $|p^i; n_1, n_2 \ldots\rangle$, in the Fock space of the a_n^i's, where n_i is the occupation number of the ith oscillator. The Hamiltonian can be written as

$$2p^+p^- = (p^i)^2 + \sum_{n=1}^{\infty}\sum_{i=1}^{d-2}\left(na_n^{i\dagger}a_n^i + \frac{n}{2}\right) \tag{5}$$

where the factor of $n/2$ represents the zero point energy of the nth oscillator.

To make sense of this formula we must interpret the divergent sum of zero point energies. A short cut to the answer (which can be more rigorously derived by lengthier methods) is to note that

$$\sum_{n=1}^{\infty} n \equiv \lim_{s\to-1}\sum_{n=1}^{\infty}\frac{1}{n^s} = \zeta(-1) = -\frac{1}{12} \tag{6}$$

where $\zeta(s)$ is the Riemann zeta function, which is defined by the series for $s > 1$ and for $s = -1$ by analytic continuation! Therefore the spectrum of the free bosonic string has states with mass M given by

$$M^2 = \sum_{n=1}^{\infty} nN_n - \frac{D-2}{24} \tag{7}$$

where N_n is the occupation number of the nth oscillator. This formula describes states that have increasing spin (as $\sum_n N_n$) and mass (as $\sum_n nN_n$), and whose degeneracy increases rapidly [as $\exp(M)$].

Let us look at the ground state and the first excited state. The ground state is the Fock vacuum, with no oscillator excitations. It has mass equal to $-(D-2)/24$, so that the lightest string state is a tachyon. What this means is that the bosonic string theory is sick, or at least the vacuum about which we are perturbing is unstable, since tachyons correspond to runaway modes. Let us ignore this problem, which in any case is avoided for superstrings, and look at the first excited string states, $a_n^{i\dagger}|\Omega\rangle$, where $|\Omega\rangle$ is the Fock space vacuum, annihilated by all the a_n^i. They have the first oscillator excited and therefore a mass of $M^2 = 1-(D-2)/24 = (26-D)/24$. There are 24 such transverse modes, $a_n^{i\dagger}|\Omega\rangle$, which transform as a vector under spatial rotations. But now we have a problem with relativistic invariance. A vector particle has $D-1$ degrees of freedom, whereas we have only $D-2$. The only exception is when the vector particle is massless, in which case the longitudinal polarization is absent (as is well known in $D = 4$ in the case of the photon). Thus, if the string theory is to be Lorentz invariant it must be that $D = 26$ so that the first excited state would be massless! Although Lorentz invariance is not manifest, it is possible to show that when $D = 26$ all the states of the free string yield representations of the Lorenz group.

This is quite remarkable. We have learned, simply by quantizing the free string, that the theory can only be formulated (without breaking Lorentz invariance or otherwise modifying the theory) in 26 space-time dimensions, and furthermore that it necessarily contains a massless vector meson. This situation is very different from that of a point particle, which can have a nonzero mass and which can be embedded in any space-time manifold of any dimensionality. It suggests that in string theory the choice of the space-time background is dynamical, so that the theory must be one of gravity, and that Yang–Mills gauge interactions are automatically induced. It is a fact of relativistic quantum mechanics that once you have a massless vector meson with nontrivial interactions (has nonvanishing couplings at zero momentum) then all of the content of local gauge theory follow (Maxwell's theory for a single massless vector meson, Yang–Mills theory for massless charged vector mesons). Charges can easily be introduced into the theory of open strings by putting them on the ends of the strings. This resembles the crude picture of a meson in QCD as a string, with the quarks at the ends of a thin flux tube of chromodynamic flux. It is of no surprise that, once one includes interactions of open strings, such a theory contains within it Yang–Mills gauge interactions.

So far we have considered open strings only. We can, and indeed must, consider closed strings as well. They are described by $x(\sigma, \tau)$, but now x is a periodic function of σ. Unlike open strings they have no geometrically invariant points on them, like the ends of open strings, to which charges can be attached. They are inherently neutral objects, that will, as we shall see, describe not gauge interactions but gravity. The closed string coordinates, in light cone gauge, obey the two-dimensional wave equation, and can be decomposed into right- and left-moving waves, $x(\sigma, \tau) = x(\sigma - \tau) + x(\sigma + \tau)$. Thus a closed string has twice as many modes, and upon quantization one finds that the mass is given by

$$\text{Mass}^2 = \sum_{n=1}^{\infty} \sum_{i=1}^{2} 4\left(n a_n^{i\dagger} a_n^i + n \tilde{a}_n^{i\dagger} \tilde{a}_n^i - 2\frac{n}{2} \right) \tag{8}$$

where $a^\dagger$ ($\tilde{a}^\dagger$) is the creation operator for right-moving (left-moving) waves, and we must impose the constraint $\sum_{n=1}^{\infty} \sum_{i=1}^{24} n a_n^{i\dagger} a_n^i = \sum_{n=1}^{\infty} \sum_{i=1}^{24} n \tilde{a}_n^{i\dagger} \tilde{a}_n^i$. Now we find that the first excited states are $a_1^{i\dagger} \tilde{a}_1^{j\dagger} |\Omega\rangle$, with $\text{Mass}^2 = (26 - D)/12$. Among these states there is a spin-two particle, which again to be Lorentz invariant must be massless. So D must equal 26, as before, and the closed string necessarily contains a massless spin-two meson—the graviton. Not surprisingly, once interactions are turned on, we will find that the theory contains within it Einstein's theory. In other words, at low energies (compared to $T \approx M^2_{\text{Planck}}$) the self interactions of the graviton, as well as its interactions with other particles, will be those of Einstein's theory.

So far we have been discussing free strings. How do we introduce interactions? In the case of pointlike theories one usually reverts at this stage to a second quantized treatment, introducing field operators that create and annihilate particles and couples them so as to produce interacting relativistic theories. One could, however, continue with the methods of first quantization, most easily by using path integral techniques. Consider again the free particle. The amplitude for this particle to go from $x^\mu(1)$ at proper time τ_1 to $x^\mu(2)$ at τ_2 is given by summing over all path histories, weighted by the exponential of the classical action

$$\langle x^\mu(2), \tau_2 | x^\mu(1), \tau_1 \rangle = \int_{x^\mu(1)}^{x^\mu(2)} Dx^\mu(\tau) \exp\left[\frac{i}{h}\int_{\tau_1}^{\tau_2} d\tau\, \mathscr{L}(\tau)\right] \tag{9}$$

where $\mathscr{L}(\tau)$ is the classical action. The integration is over all trajectories up to reparametrizations. A similar expression can be written for the free string theory. To introduce interactions we must consider the motion of two particles and modify the rules whenever their trajectories cross, allowing for the particles to combine to form a third particle, or for a single particle to split into two. This is certainly not a unique, or geometrically inspired, procedure and introduces much arbitrariness into the theory of point particles. In the case of strings, however, there is a natural, geometric way of introducing interactions that is unique. The motion of a single closed string is gotten by summing over all path histories, each one of which looks like a cylinder in σ, τ space. If we add to the cylinder a handle, then the space-time process described by such a world sheet looks like a closed string which propagates, then splits into two closed strings, which in turn propagate and then rejoin—namely, a self-energy (one-loop) correction to the string propagator.

To see this we have to follow the motion as a function of τ, and then at a specific value of τ the strings split or rejoin. On the other hand, if we look at the surface as a whole there is no place where anything singular happens (at least for the *Euclidean* surface we get by analytically continuing to complex τ), no place where an interaction is introduced. Thus the complete amplitude in string theory can be written schematically as

$$\sum_{\text{topologies}} \int Dx^\mu(\sigma, \tau) \exp\left[\frac{i}{h}\int_{\tau_1}^{\tau_2} d\tau \int d\sigma \mathscr{L}[x^\mu(\sigma, \tau)]\right] \tag{10}$$

where we sum over all topologies of the two-dimensional world sheets connecting our initial and final string configurations.

Such a method for introducing interactions clearly preserves the geometrical invariance of the theory, and thus the ensuing gauge symmetries.

Doing the above integrals is somewhat nontrivial; after all, these are functional integrals over two-dimensional "fields" $x^\mu(\sigma, \tau)$. Each integral in the above sum is equivalent to constructing a two-dimensional field theory on a surface of a given topology. In general this would be a formidable task. In the case of the bosonic string in flat (26-dimensional) space-time the situation is not so bad—in fact we saw, as in (3), that with the appropriate gauge choice the theory is a free field theory. The path integrals can then be done quite explicitly, the only complications arising from the need to factor out correctly the reparametrization group, and from the fact that one is working on a surface with complicated topology. In the last two years the methods for doing these integrals have been developed to an extraordinary degree, using high-powered methods of analytic geometry and complex analysis.

String theories are inherently theories of gravity. Unlike ordinary quantum field theory, however, we do not have the option of turning off gravity. The gravitational, or closed string, sector of the theory must always be present for consistency, even if one starts by considering only open strings, since these can join at their ends to form closed strings. One could even imagine discovering the graviton in the attempt to construct string theories of matter. In fact this was the course of events for the dual resonance models where the graviton (then called the Pomeron) was discovered as a bound state of open strings. In this theory the ordinary gauge couplings, like the fine structure constant, are proportional to the gravitational coupling, so if one turns off gravity one turns off everything. The heterotic string, which is a purely closed string theory, can be regarded as a gravitational theory that produces gauge interactions by a stringy version of the Kaluza-Klein mechanism.

Most exciting is that string theories provide for the time a consistent, even finite, theory of gravity. The problem of ultraviolet infinities is bypassed in string theories, which simply contain no short-distance singularities. This is not too surprising considering the extended nature of strings, which softens their interactions. Alternatively one notes that interactions are introduced into string theory by allowing the string coordinates, which are two-dimensional fields, to propagate on world sheets with nontrivial topology that describe strings splitting and joining. From this first quantized point of view one does not introduce an interaction at all, one just adds handles or holes to the world sheet of the free string. As long as reparametrization invariance is maintained there are no possible counterterms. In fact all the infinities that have ever appeared in string theories can be traced to infrared divergences that are a consequence of vacuum instability. These arise since all string theories contain a massless partner of the graviton called the dilaton. If one constructs a string theory about a trial vacuum state in which the dilaton has a nonvanishing vacuum expectation value, then infrared infinities will occur due to massless dilaton tadpoles. These divergences, however, are just a sign of the instability of the original trial vacuum. This is the source of the divergences that occur in one loop diagrams in the old bosonic string theories (the Veneziano model). Superstring theories have vanishing dilaton tadpoles, at least to one-loop order. Therefore both the superstring and the heterotic string are explicitly finite to one loop order and there are strong arguments that this persists to all orders!

String theories, as befits unified theories of physics, are incredibly unique. In principle they contain no freely adjustable parameters and all physical quantities should be calculable in terms of h, c, and m_{Planck}. In practice we are not yet in a position to exploit this enormous predictive power. The fine structure constant α, for example, appears in the theory in the form $\alpha \exp(-D)$, where D is the aforementioned dilaton field. Now, the value of this field is undetermined to all orders in perturbation theory (it has a "flat potential"). Thus we are free to choose its value, thereby choosing one of an infinite number of degenerate vacuum states, and thus to adjust α as desired. Ultimately we might believe that string dynamics will determine the value of D uniquely, presumably by a nonperturbative mechanism, and thereby eliminate the nonuniqueness of the choice of vacuum state. In that case all dimensionless parameters will be calculable. Even more, string theories determine the gauge group of the world (to be $E_8 \times E_8$ or possibly SO_{32}) and fix the number of space-time dimensions to be ten. This might appear to be disastrous, since the world we observe about us has only four dimensions, and a recognizable gauge group of $SU_3 \times SU_2 \times U_1$. However, as we shall see below the heterotic string theory can have phenomenologically attractive solutions, which could well describe the real world.

The number of consistent string theories is extremely small, the number of phenomenologically attractive theories even smaller. First, there are the closed superstrings, of which there are two consistent versions. These are theories that contain only closed strings that have no ends to which to attach charges and are thus inherently neutral objects. At low energies, compared to the mass scale of the theory, which we can identify as the Planck mass, we only see the massless states of the theory, which are those of ten-dimensional supergravity. One version of this theory is nonchiral and of no interest since it could never reproduce the observed chiral nature of low-energy physics. The other version is chiral. One might then worry that it might suffer from anomalies, which is indeed the fate of almost all chiral supergravity theories in ten dimensions. Such anomalies are ubiquitous features of theories containing chiral fermions. They are disastrous since they mean that quantum corrections spoil the gauge invariance of the theory. In the standard model, in fact, the would-be anomalies cancel only as a consequence of a conspiracy between quarks and leptons. This conspiracy is one of the strongest arguments that nature wants to be chirally asymmetric and goes to some length to achieve this (after all, one could eliminate all possible anomalies by simply imposing manifest chiral symmetry). It is also a strong indication that quarks and leptons are highly correlated, and all by itself suggests unification of the strong and electroweak interactions.

Remarkably, the particular supergravity theory contained within the chiral superstring is the unique anomaly-free pure supergravity theory in ten dimensions. Although consistent, it contains no gauge interactions in ten dimensions and could only produce such as a consequence of compactification. This approach raises the same problems of reproducing chiral fermions that plagued field theoretic Kaluza–Klein models and has not attracted much attention.

Open string theories, on the other hand, allow the introduction of gauge groups by the time-honored method of attaching charges to the ends of the strings. String theories of this type can be constructed which yield, at low energies, $N = 1$ supergravity with any Yang–Mills group. These, in addition to being somewhat arbitrary, were suspected to be anomalous. The discovery by Green and Schwarz that, for a particular gauge group SO_{32} the would-be anomalies cancel, greatly increased the phenomenological prospects of unified string theories.

The anomaly cancellation mechanism of Green and Schwarz also provided the motivation that led to the discovery of a new string theory, the *heterotic* string. This is a theory of closed strings, which generates nonetheless gauge interactions. Quite different from the ad hoc procedure of attaching charges to string end points this mechanism determines the gauge group uniquely. There are two manifestations, which we now understand as different states of the same theory, realized as $E_8 \times E_8$ or spin-$32/Z_2$ gauge symmetry groups. The $E_8 \times E_8$ version of this theory offers the best phenomenological prospects for reproducing the real world. In fact the group E_8 was explored seriously as a GUT group by theorists who extrapolated upward from the standard model, so one might hope to be able to proceed in the opposite direction.

In the next section I shall discuss the construction of the heterotic string theory in detail. For now let us discuss its possible connection to the real world.

In order to make contact between string theories and the real world one is faced with a formidable task. These theories are formulated in ten flat space-time dimensions, have no candidates for fermionic matter multiplets, are supersymmetric, and contain an unbroken large gauge group—say $E_8 \times E_8$. These are not characteristic features of the physics that we observe at energies below 1 TeV. If the theory is to describe the real world one must understand how six of the spatial dimensions compactify to a small manifold leaving four flat dimensions, how the gauge group is broken to $SU_3 \times SU_2 \times$

U_1, how supersymmetry is broken, how families of light quarks and leptons emerge, etc. Much of the recent excitement concerning string theories has been generated by the discovery of a host of mechanisms, due to the work of Witten and of Candelas, Horowitz, and Strominger, and of Dine, Kaplonovsky, Nappi, Seiberg, Rohm, Breit, Ovrut, Segre, and others, which indicate how all of this could occur. The resulting phenomenology, in the case of the $E_8 \times E_8$ heterotic string theory, is quite promising.

1.1. Compactification

The first issue to be addressed is that of the compactification of six of the dimensions of space. The heterotic string, as described above, was formulated in ten-dimensional flat space-time. This, however, is not necessary. Since the theory contains gravity the issue of which space-time the string can be embedded in is one of string dynamics. That the theory can consistently be constructed in perturbation theory about flat space is equivalent to the statement that ten-dimensional Minkowski space-time is a solution of the classical string equations of motion. In quantum mechanics such a solution yields the background expectation values of the quantum degrees of freedom. We can then ask, are there other solutions of the string equations of motion that describe the string embedded in, say, four-dimensional Minkowski space-time times a small compact six-dimensional manifold?

At the moment we do not possess the full string functional equations of motion; however, one can attack this problem in an indirect fashion. One method is to deduce from the scattering amplitudes that describe the fluctuations of the string in ten-dimensional Minkowski space an effective Lagrangian for local fields that describe the string modes. Restricting one's attention to the massless modes, the resulting Lagrangian yields equations that reduce to Einstein's equations at low energies, and can be explored for compactified solutions. Another method is to proceed directly to construct the first quantized string about a trial vacuum in which the metric (as well as other string modes) have assumed background values. In the third lecture I shall describe in some detail these two approaches and the results that have been achieved to date. For today let us discuss the solutions.

Remarkably, there do exist a very large class of conformally invariant supersymmetric σ models that yield solutions of the string classical equations of motion to all orders in perturbation theory, and describe the compactification of ten dimensions to a product space of four-dimensional Minkowski space times a compact internal six-dimensional manifold. These compact manifolds are rather exotic mathematical constructions. They are Kahler manifolds (which means that one can define, as in the case of the complex plane, a global complex set of coordinates), admit a Ricci flat metric (which means that they are solutions of Einstein's equations, $R_{ab} = 0$), and they have SU_3 holonomy (which means that when we parallel transport a vector around a closed curve on the six-dimensional manifold it undergoes a rotation by an SU_3 transformation). Such spaces are called "Calabi–Yau" manifolds. There are many (of order tens of thousands) such manifolds and each has, in general, many free parameters (moduli). These moduli determine the shape and size of the compact space. This is an indication of the enormous vacuum degeneracy of the string theory, at least when treated perturbatively, and leads (at the present stage of our understanding) to many free parameters. This abundance of riches should not displease us. We first would like to know whether there are any solutions of the theory that resemble the real world; later we can try to understand why the dynamics picks out a particular solution. It is very pleasing that there are many solutions for which four of the dimensions are flat, namely, the cosmological constant vanishes, whereas the other six are curled up. So even if we do not know why six dimensions of space necessarily curl up, we learn that they could do so.

1.2. The Low-Energy Gauge Group

We now have to see whether we can find solutions in which the observed gauge group is not $E_8 \times E_8$, but rather that of the standard model. How does this immense group get broken? The first breaking that occurs is a consequence of the space-time compactification. In the case of the heterotic string it is not sufficient to simply embed the string in a Calabi–Yau manifold. One must also turn on an SU_3 subgroup of the $E_8 \times E_8$ gauge group of the string. This is because the internal degrees of freedom of the heterotic string consist of right-moving fermions, which feel the curvature of space-time, and left-moving coordinates, which know nothing of the space-time curvature but are sensitive to background gauge fields. Unless there is a relation between the curvature of space and the curvature (field strength) of the gauge group there is a right–left mismatch which gives rise to anomalies. Alternatively, one can explore the effective Lagrangian of the massless modes of the string theory. One finds that the equation of motion for the dilaton field, (Φ), is of the form

$$\Delta\Phi \propto R_{\alpha\beta\gamma\delta}R^{\alpha\beta\gamma\delta} - F^{cd}_{\alpha\beta}F^{\alpha\beta}_{cd} \tag{11}$$

where $R_{\alpha\beta\gamma\delta}$ is the Riemann tensor, and $F^{cd}_{\alpha\beta}$ is the gauge field strength, and Δ is the Laplacian of the internal manifold. For a solution with $\Phi = \text{const}$ we require a cancelation between the curvature of the manifold and that of gauge space. The easiest way to satisfy this is to identify the space-time curvature with the gauge curvature (embed the spin connection in the gauge group). One does this by turning on background gauge fields in an SU_3

subgroup of one of the E_8's (i.e., we set $F^{cd}_{\alpha\beta} = R_{\alpha\beta cd}$ using the standard imbedding of SU_3 in O_6). This has the nice feature of breaking E_8 down to E_6, which is a much better GUT group. This feature of the heterotic string, namely, the tight relation between the curvature of the internal (gauge) space and that of space-time, is much more general than these particular compactifications. More generally, we can imagine the background gauge fields being in other subgroups of O_6. This would then allow the possibility of other unbroken gauge groups, O_{10} or SU_5.

This leaves us with an unbroken $E_6 \times E'_8$, which is still much too big. In older GUT approaches one would at this stage introduce by hand a bunch of scalar (Higgs) particles, and arrange the scalar potential so that enough of the scalars get expectation values to break the large symmetry. Here we are not allowed to do this; the theory contains all that there is and we cannot add anything. However, there is a very nice and natural mechanism for the breaking of E_6 down to the observed low-energy gauge group. This mechanism is more general than string theory; it is a possible symmetry-breaking mechanism for any theory that contains compactified dimensions. The point is that the internal six-dimensional manifold (which we shall call K) is, in general, multiply connected. This means that K is full of holes, and we can draw closed paths on K that go around these holes and cannot be continuously shrunk to nothing. Now, if K/Z is multiply connected one can allow flux of the unbroken E_6 (or of the E'_8 for that matter) to run through it, with no change in the vacuum energy. The net effect is that when we go around a hole in the manifold, through which some flux runs, we must perform a nontrivial gauge transformation on the charged degrees of freedom. So, even though the flux is running through holes where there is no manifold it has a physical effect. In the case of electromagnetism this effect is known as the Bohm–Aharonov effect, whereby a flux of magnetic field can affect charged particles that are restricted to a multiply connected region of space that has vanishing field strength.

Another way of describing this is to say that we have nonvanishing Wilson loops around the noncontractible paths on K. These noncontractible Wilson loops act very much like Higgs bosons, breaking E_6 down to the largest subgroup that commutes with all of them. By this mechanism one can, without generating a cosmological constant, find vacua whose unbroken low-energy gauge group is, say, $SU_3 \times SU_2 \times U_1 \times$ (typically, an extra U_1 or two). This method of symmetry breaking is quite different in some respects from that induced by explicit Higgs bosons. In particular E_6 is in no sense an approximate symmetry of the theory. It is inherently broken at the compactification scale, and above this scale we have the full E_8 symmetry. This, as I shall argue below, is actually quite nice, since it means that various parameters of the theory are not related by E_6 symmetry at the compactification scale.

On the other hand the gauge couplings of the unbroken gauge groups (SU_3, SU_2, and U_1) are all equal at the compactification scale. This is sufficient for us to attempt to repeat the calculation of Georgi, Quinn, and Weinberg of the Weinberg angle. Recall that this angle is determined by the ratios of the couplings at low energies; these in turn are equal at the unification scale (in our theory as well), but renormalize differently as one lowers the energies. The precise way in which the different couplings run depends on the matter content of the theory. In fact it can be used to place strong constraints on the matter content, as we shall discuss below.

Is this sufficient? What about the other E_8'? It too could be broken by the above mechanism, but it is really not necessary. An unbroken E_8' is quite unobservable at low energy unless there exist light matter fields with E_8' quantum numbers. As we shall see below there are no such particles for these solutions. In that case E_8' physics is that of a QCD-like theory with no quarks. The mass scale of such a theory would be very large since the effective coupling grows faster than the QCD coupling as we go to lower energies. This is because the β function that controls the running coupling is proportional to the Casimir of the gauge group, and this is much larger for E_8 than for SU_3. Thus we would expect to have very heavy E_8' glueballs, and no significant effects on low-energy physics. Actually, such a strong E_8' gauge group might be useful, as has been conjectured, as a way of breaking supersymmetry.

In conclusion it is very pleasing that there are many solutions that have the standard model as the observable low-energy gauge group (plus perhaps some extra U_1's, which would be welcome, if there, as a way of testing the theory).

1.3. The Matter Content

These Calabi-Yau compactifications produce for each manifold K a consistent string vacuum, for which the gauge group is no larger than $E_6 \times E_8$. Now we must address the issue of the *matter* content of the theory. This is determined by the massless spectrum of the theory. We must determine the quantum numbers of the massless particles and see whether they can correspond to the observed families of quarks and leptons. After Kaluza-Klein compactification the spectrum of massless chiral fermions is determined by the zero modes of the Dirac operator on the internal space. This is because the Dirac equation for a fermion reads $\gamma^\mu D_\mu \Psi(x^\mu) = 0$, where μ runs over the ten space-time indices. If we have compactified six of the spatial dimensions (X^n; $n = 4, 5, 6, 7, 8, 9$) it is convenient to expand Ψ in terms of a basis of functions, $\Phi_k(X^n)$, on the compact manifold K, which are eigenfunctions of the internal Dirac operator $D_{\text{int}} = \sum_{n=1}^{6} \gamma^n D_n$, i.e., $D_{\text{int}}\Phi_k = \lambda_k \Phi_k$. In this basis we can write $\Psi(x^\mu) = \sum_k \Phi_k(X^n)\Psi_k(x^\alpha)$, where x^α; $\alpha = 0, 1, 2, 3$ runs over ordinary Minkowski space. In that case the four-dimensional fields, $\Psi(z^\alpha)$, satisfy the four-dimensional Dirac equation

$$[\gamma^\alpha D_\alpha + \lambda_k]\Psi_k(x^\alpha) = 0 \tag{12}$$

where the eigenvalue, λ_k, of the internal Dirac operator emerges as the four-dimensional mass. For every nontrivial solution of the Dirac equation with $\lambda_k = 0$ in the internal space there will appear a massless fermion in four dimensions.

Now it might appear to be a difficult task to decide how many solutions there are to this equation. The number might depend on grubby details of the internal manifold. This is not the case; one can determine the number of massless fermions by topological arguments! To see this let us recall the chiral properties of the Dirac equation. Remember that in any even number of dimensions, d, there is a matrix γ^{d+1}, analogous to γ^5 in four dimensions, that anticommutes with all the γ matrices, $\{\gamma^{d+1}, \gamma^\mu\} = 0$; $\mu = 1, \ldots, d$. Now consider the Dirac equation

$$[\gamma^\mu D_\mu + \lambda]\Psi(x^\mu) = 0 \tag{13}$$

If we have a solution Ψ of this equation, with eigenvalue $\lambda \neq 0$, then $\gamma^{d+1}\Psi$ is also a solution, but with eigenvalue $-\lambda$. Thus the solutions with nonvanishing eigenvalues necessarily come in pairs, with equal and opposite values. This need not be the case when $\lambda = 0$, in which case the solutions can be chosen to be of definite *chirality*, $\gamma^{d+1}\Psi = \pm\Psi$, and need not be paired. Now consider changing the form of the internal manifold K slightly. This will change the Dirac operator and its spectrum. In particular we might imagine that the number of zero eigenvalues could change abruptly. Since we cannot lose a zero eigenvalue by a continuous change in K, it must turn into a nonzero eigenvalue. However, the nonzero eigenvalues come in pairs ($\pm\lambda$). This means that two zero eigenvalues must disappear at the same time. Furthermore, they must have opposite chiralities so that we can form the appropriate linear combinations. We therefore learn from this simple argument that *under continuous changes in the manifold the number of positive chirality, N_+, minus the number of negative chirality zero modes, N_-, is unchanged.* This means that $N_z = N_+ - N_-$ (the *index* of the Dirac operator) is a topological property of K, in other words a property that is invariant under continuous deformations of the manifold.

This is a very nice feature of the zero modes of the Dirac operator. It means that N_z can be calculated in terms of topological properties of K alone. Note that this argument only determines N_z and not N_+ and N_- separately, but that is fine since we are actually only interested in N_z. This is because one positive and one negative chirality fermion can always combine and become a massive fermion (recall that a Dirac mass term does not commute with γ^{d+1} and mixes chiralities). Therefore we would expect, that if, say, $N_z \geq 0$ then N_- fermions will acquire a mass (in string theory typically of order M_{Planck}) and decouple, leaving us with N_z massless fermions. (Actually, in string theory one can also determine the numbers N_+ and N_- separately by topological arguments.)

Now what are the quantum numbers of possible massless fermions? Since the fermions are originally all in the adjoint representation of $E_8 \times E_8$, the $\mathbf{248} \oplus \mathbf{248}$ representation, the massless fermions that emerge after compactification must appear in decompositions of this representation. Now the adjoint representation of E_8 decomposes as $(\mathbf{1}, \mathbf{78}) \oplus (\mathbf{3}, \mathbf{27}) \oplus (\bar{\mathbf{3}}, \bar{\mathbf{27}}) \oplus (\mathbf{8}, \mathbf{1})$ under $SU_3 \times E_6$. We are interested in the $(\mathbf{27})$, which contains one standard family of quarks and leptons. To see the familiar quarks and leptons let us decompose E_6 under its maximal subgroup $SU(3)_{\text{color}} \times SU(3)_{\text{left}} \times SU(3)_{\text{right}}$. Here we have identified these groups as color and as left- and right-handed electroweak interactions, respectively. Then the $\mathbf{27}$ can be decomposed as follows:

$$\mathbf{27} \to (\bar{\mathbf{3}}, \mathbf{3}, \mathbf{1}) + (\mathbf{3}, \mathbf{1}, \mathbf{3}) + (\mathbf{1}, \bar{\mathbf{3}}, \bar{\mathbf{3}}) \tag{14}$$

A standard family of quarks and leptons can be easily accommodated in this representation. The up and down quarks, together with a new, charge $-1/3$ quark (called g) form an $SU(3)_{\text{left}}$ multiplet and their antiparticles an $SU(3)_{\text{right}}$ multiplet. The $(\mathbf{1}, \bar{\mathbf{3}}, \mathbf{3})$ contains the electron and the electron neutrino, as well as some new particles, all in all

$$\begin{pmatrix} H_1^u & H_1^d & e^- \\ H_2^u & H_2^d & \nu \\ e^+ & \bar{\nu} & S_2 \end{pmatrix}$$

Of course, one must suppose that the new particles in this multiplet get large masses, which explains why they have not yet been observed. Note that the particles $H_{1,2}^{u,d}$ have the quantum numbers of Higgs particles, so that their scalar partners could be responsible for the breaking of $SU(2)_L \times U(1)_{EM}$ to $U(1)_{EM}$ at low energy.

Since, for the heterotic string, the gauge and spin connections are forced to be equal, one can count the number of chiral fermions by geometrical arguments. This works in the following way. Let N_{27}^L (N_{27}^R) be the number of left(right)-handed massless fermion multiplets transforming as a $\mathbf{27}$ under E_6 and let $N = N_{27}^L - N_{27}^R$. The number of generations is just equal to $|N|$. Now fermions transforming in the $\mathbf{27}$ subgroup of E_8 also transform in the $\mathbf{3}$ of $SU(3)$. Thus N is just the index of the Dirac operator on the internal manifold K acting on spinors in the **3** representation of $SU(3)$. But the $SU(3)$ gauge connection (vector potential) is equal to the spin connection, so that the index can only depend on the topology of K. In fact it is equal to one half of the Euler number, χ, of K

$$|N| = \tfrac{1}{2}|\chi| \tag{15}$$

So the number of generations is equal to half the Euler character of the manifold (which, if the manifold were two dimensional, would simply count the number of handles). For most Calabi–Yau manifolds the Euler character turns out to be quite large. There is, however, a way of reducing it. If there exists a discrete symmetry group, Z, which acts freely on K (which means that the group has no fixed points), one can consider the smaller manifold K/Z (in other words, points on K that are related by a symmetry transformation of Z are identified). If Z acts freely on K this produces a new manifold whose volume is reduced by the dimension of Z. If Z has fixed points one gets a manifold with singular points at the fixed points—an orbifold) On K/Z the Euler character is reduced by the dimension of Z. By this trick, and after some searching, manifolds have been constructed with 1, 2, 3, 4, . . . , 200 generations.

How many generations can there be? So far we have observed three generations of quarks and leptons, but clearly there might be many more if they are sufficiently heavy. There are, however, indirect ways of putting bounds on the number of generations. For example, it is possible to argue on the basis of standard big bang cosmology that there cannot be more than ≈ 4 massless neutrinos. More neutrinos would seriously affect the calculation of helium production, one of the most successful of the predictions of standard cosmology. A direct bound on the number of light neutrinos will shortly be provided by precision measurements of the W and Z masses. From a theoretical point of view the number of generations in a unified theory is bounded, if we are to come up with a successful prediction of the Weinberg angle. The point is that the ratio of the various couplings of the subgroups of a unified gauge group are renormalized separately once the full symmetry is broken. In string theory, for example, they are all equal at the compactification scale, but differ at low energies where their ratio determines the weak mixing angle. The number of quark and lepton generations affects this differential renormalization, and so affects the calculation of the Weinberg angle. It seems that to be realistic we must restrict to manifolds with three, or perhaps four, generations.

Once again we find an enormous number of possible solutions. It is extremely pleasing that the theory automatically produces chiral fermions. This was not the case for other Kaluza–Klein theories. It is even more pleasing they can produce families of quarks and leptons with the right quantum numbers.

1.4. Further Symmetry Breaking

We have shown, so far, that there exist solutions of the heterotic string theory in four-dimensional Minkowski space (times a small compact manifold), that the large gauge symmetry can be broken down to the standard model symmetry group, and that there are solutions with roughly the right number and kinds of massless fermions to be identified with the observed quarks and leptons. However, there is still much to be done before we can directly compare with the real world.

First of all, all of these solutions have exact $N = 1$ supersymmetry. This is good and bad. It is good because we need to have supersymmetry survive all the way down to low (≈ 1 TeV) energy if we want supersymmetry to solve the *hierarchy* problem by protecting the masslessness of the Higgs particles. It is bad because supersymmetry is clearly not an exact symmetry of nature; in fact it is so badly broken that no sign of its existence has yet appeared. It is therefore necessary to break the remaining $N = 1$ supersymmetry. However, one must be careful. There are easy ways of breaking the supersymmetry, even in perturbation theory. For example, one can with toroidal compactifications introduce twisted boundary conditions that violate supersymmetry. However, this has bad consequences. First, it breaks supersymmetry at the compactification scale, which is probably too high. Second, it produces at one loop order a cosmological constant and destabilizes the vacuum. The generation of a cosmological constant is always a potential problem once supersymmetry is broken. It does not seem likely that perturbative supersymmetry breaking could exist without generating an intolerably big cosmological constant. So we must contemplate nonperturbative mechanisms.

For this purpose the extra E_8 gauge group might be useful. Below the compactification scale it yields a strong, confining gauge theory like QCD, but without light matter fields. In general this sector would be totally unobservable to us, consisting of very heavy glueballs, which would only interact with our sector with gravitational strength at low energies. However, there could very well exist in this sector a gluino condensate which can serve as source for supersymmetry breaking. This possibility has been considered, but so far does not seem to work. What goes wrong? The problem appears to be that once supersymmetry is broken the dilaton expectation value can be dynamically fixed, in other words there is now a nonflat dilaton potential. Now there is always one stable point for the dilaton, namely, where its expectation value, ϕ, blows up. Since the couplings scale as $e^{-\phi}$ this means that all couplings vanish and the theory is free. When supersymmetry is broken, at least by the mechanism discussed above, the theory tends to relax to a free theory, or, equally bad, to ten-dimensional flat space.

This problem—how to break supersymmetry without producing a cosmological constant—is the major obstacle to relating the heterotic string to the real world. It is likely that the solution must await the development of the theory to the point where nonperturbative issues can be addressed and answered. If we were to succeed in breaking supersymmetry at low energies, then the remaining ingredients for a successful description of the real world are already in place. There exists a natural reason for the existence of massless Higgs bosons which are weak isospin doublets (and could be responsible for the electroweak breaking at a TeV), without accompanying color triplets.

Thus the heterotic string theory appears to contain, in a rather natural context, many of the ingredients necessary to produce the observed low-energy physics. I do not mean to suggest that there are not many problems and unexplained mysteries. In addition to those discussed above there exists the danger (common to many grand unified models, especially supersymmetric ones) of too rapid proton decay; there is no deep understanding of why the cosmological constant, so far zero, remains zero. Nonetheless the early successes are very reassuring and they give one the feeling that there are no insuperable obstacles to deriving all of low-energy physics from the $E_8 \times E_8$ heterotic string theory.

DISCUSSION

R. JACKIW: You stated that Calabi–Yau manifolds are solutions of string theory to all orders in perturbation theory. Of which perturbation theory are you talking?

D. GROSS: There are two kinds of perturbation theory that arise in string theory: perturbation theory of the classical equations of motion and quantum perturbation theory. I was discussing the first kind since we are first interested in finding solutions of the classical field equations. Here the expansion parameter is the deviation of the metric from flat space, as measured, say, by the ratio of the Planck length to the radius of the internal space. This parameter is proportional to the coupling constant of the two-dimensional nonlinear σ model whose β function one constructs, perturbatively, to find conformally invariant field theories, i.e., classical string solutions. It has also been argued, quite persuasively, that the quantum perturbation theory about such a classical solution will not exhibit any instabilities, so that these classical solutions can be consistent starting points for a quantum perturbation theory.

S. ADLER: Don't you require a cosmological constant in order to have an inflationary cosmology?

GROSS: I have been discussing the particle physics vacuum and not a cosmolgical solution of the theory. Normally we distinguish these. To do cosmology we start the universe out in some initial state, say a hot dense universe, and watch it expand. As time goes on it resembles more and more the stationary state that we identify with the particle physics vacuum. It is this stationary state that is the normal concern of particle physics and must have zero, or very small, cosmological constant. This does not contradict the inflationary models, since they are referring to cosmological solutions in which the universe starts out in an excited state. In other words the ground state of the Hamiltonian has zero energy density, but one might put the universe into an excited state with nonzero energy density.

Let me note that no one has yet discussed cosmology in the context of string theory in a satisfactory fashion, and surprises might very well arise, especially at early, Planckian, times. In particular one might wonder whether string theory could be so powerful as to pick out a unique cosmological solution, and thereby determine the initial conditions of the universe.

2. HETEROTIC STRING THEORY

In this section I discuss some aspects of the heterotic string theory, in particular the special stringy Kaluza–Klein mechanism that produces gauge interactions from this theory of closed strings.

Previously known string theories are the bosonic theory in 26 dimensions (the Veneziano model) and the fermionic, superstring theory in 10 dimensions (an outgrowth of the Ramond–Neveu–Schwarz string). The new heterotic string theory is constructed as a chiral hybrid of these. The basic idea underlying the construction of the heterotic string is the separation, for orientable closed strings, between right- and left-moving modes. The physical degrees of freedom of a closed string (which can be regarded as two-dimensional massless fields), can be decomposed into right and left movers, i.e., functions of $\tau - \sigma$ and $\tau + \sigma$, respectively. A right-moving wave will travel uninterupted and will never turn into a left-moving wave. (We need to consider orientable strings so that one can distinguish right from left.) For open strings the boundary conditions mix left and right movers, since a right-moving wave is reflected back from the end as a left-moving wave. However, if we consider only closed strings, then the right and left movers never mix. All observables can be decomposed into operators acting separately on the left- and right-handed Hilbert space of the string states. There is only one constraint that ties the two halfs of the closed string together to form a geometrical object, namely, that the right- and left-handed number operators are equal, $N_R = N_L$. Apart from this constraint everything else factorizes and we can write Fock space states as a direct product $|\rangle_R \times |\rangle_L$.

This separation is maintained even in the presence of string interactions, as long as we allow only orientable world sheets on which a handedness can be defined. This is because the interactions between closed strings are constructed, order by order in perturbation theory, by simply modifying the topology of the world sheet on which the strings propagate. In terms of the first quantized two-dimensional theory no interaction is thereby introduced; the right and left movers still propagate freely and independently as massless fields.

Thus, there is in principle no obstacle to constructing the right and left moving sectors of a closed string in a different fashion, as long as each sector is separately consistent, and together can be regarded as a string embedded in ordinary space-time. This is the idea behind the construction of the heterotic string, which combines the right movers of the fermionic superstring with the left movers of the bosonic string. The bosonic theory contains, as we have seen, a tachyon. This, however, is not a sign of *inconsistency*, but rather a sign of *instability*. In any case the tachyon does not show up in the heterotic theory.

We shall now discuss the quantization of the heterotic string, with emphasis on the new degrees of freedom. We shall work in light-cone gauge as before. The physical degrees of freedom of the right-moving sector of the fermionic superstring consists of eight transverse coordinates $x^i(\tau - \sigma) \times (i = 1, \ldots, 8)$ and eight Majorana–Weyl fermionic coordinates $S^a(\tau - \sigma)$. It is the symmetry between these eight fermionic and eight bosonic modes that leads to supersymmetry. These have the normal mode expansions

$$x^i(\tau - \sigma) = \tfrac{1}{2}x^i + \tfrac{1}{2}p^i(\tau - \sigma) + \frac{i}{2}\sum_{n\neq 0} \frac{\alpha^i_n}{n} e^{-2in(\tau-\sigma)} \tag{16}$$

$$S^a(\tau - \sigma) = \sum_{n=-\infty}^{\infty} S^a_n e^{-2in(\tau-\sigma)} \tag{17}$$

The physical degrees of freedom of the left-moving sector of the bosonic string consist of 24 transverse coordinates, which we divide into two classes, $x^i(\tau + \sigma)$ and $x^I(\tau + \sigma)$ $(i = 1, \ldots, 8, I = 1, \ldots, 16)$, with the expansion

$$\begin{aligned} x^i(\tau + \sigma) &= \tfrac{1}{2}x^i + \tfrac{1}{2}p^i(\tau + \sigma) + \frac{i}{2}\sum_{n\neq 0} \frac{\tilde{\alpha}^i_n}{n} e^{-2in(\tau+\sigma)} \\ x^I(\tau + \sigma) &= x^I + P^I(\tau + \sigma) + \frac{i}{2}\sum_{n\neq 0} \frac{\tilde{\alpha}^I_n}{n} e^{-2in(\tau+\sigma)} \end{aligned} \tag{18}$$

Together these comprise the physical degrees of freedom of the heterotic string. The eight transverse right and left movers combine with the longitudinal coordinates x^+ and x^- to describe the position of the string embedded in ten-dimensional space. The extra fermionic and bosonic degrees of freedom parametrize an internal space. The ten fermions are responsible for $N = 1$ supersymmetry, whereas the sixteen bosonic coordinates will provide the arena for the gauge symmetries.

The α_n's (S_n^a's) are standard creation and annihilation operators for the bosonic (fermionic) modes of vibration of the string, satisfying

$$\begin{gathered}[\alpha_n^i, \alpha_m^j] = [\tilde{\alpha}_n^i, \tilde{\alpha}_m^j] = n\delta_{n+m,0}\delta^{ij}, \qquad [\alpha_n^i, \tilde{\alpha}_m^j] = 0 \\ [\tilde{\alpha}_n^I, \tilde{\alpha}_m^J] = n\delta_{n+m,0}\delta^{IJ}, \qquad [S_m^a, S_n^b] = [\gamma^+(1-\gamma^{11})]^{ab}\delta_{n+m,0}\end{gathered} \tag{19}$$

The only subtle feature entering into the quantization of the heterotic string is the treatment of the center-of-mass position and momentum of the left-moving coordinates, X^I and P^I. These are unusual since the constraint (a second class constraint) that $X^I(\tau+\sigma)$ satisfy, $(\partial_\tau - \partial_\sigma)X^I = 0$, is not consistent with the canonical commutator of X^I and P^I. We must, therefore, modify this commutator à la Dirac. This leads to

$$[X^I, P^J] = \frac{i}{2}\delta^{IJ} \tag{20}$$

which differs by a factor of 1/2 from the more standard $[x^i, p^i] = i\delta^{ij}$. A quick way of seeing this factor of 1/2 is to note that if we had both right and left movers, $x^R(\tau+\sigma) = x^R + P^R(\tau+\sigma) + \cdots$; $x^L(\tau-\sigma) = x^L + P^L(\tau-\sigma) + \cdots$, then in order to assure that $[x^R + x^L, p^R + p^L] = i$ and that x^R (x^L) commute with p^L (p^R), we must have $[x^L, p^L] = [x^R, p^R] = i/2$.

The mass spectrum of the heterotic string is determined by the equation

$$\frac{\text{Mass}^2}{4} = N + \tilde{N} - 1 + \frac{1}{2}\sum_{I=1}^{16}(P^I)^2 \tag{21}$$

where N ($\tilde{N}$) is the number operator of the right (left) movers,

$$\begin{aligned} N &= \sum_{n=1}^{\infty}[\alpha_{-n}^i\alpha_n^i + \tfrac{1}{2}nS_{-n}\gamma^- S_n] \\ \tilde{N} &= \sum_{n=1}^{\infty}[\tilde{\alpha}_{-n}^i\tilde{\alpha}_n^i + \tilde{\alpha}_{-n}^I\tilde{\alpha}_n^I]\end{aligned} \tag{22}$$

The subtraction of -1 in (21) is due, as discussed before, to the normal ordering of $\tilde{N}$, which is absent in the case of the right movers due to the supersymmetric boson fermion cancellation of the zero-point energy. Finally, the factor of $(P^I)^2$ comes from the internal momentum of the left movers. These number operators must be related, in fact must satisfy

$$N = \tilde{N} - 1 + \frac{1}{2}\sum_{I=1}^{16}(P^I)^2 \tag{23}$$

This constraint has a simple geometrical interpretation, namely, that the origin, $\sigma = 0$ is not a distinguished point on the string. The unitary operator that shifts this origin by an amount Δ is just $U(\Delta) = \exp\{2i\Delta[N - \tilde{N} - 1 + \frac{1}{2}\sum_{I=1}^{16}(P^I)^2]\}$, since one can easily show that $U(\Delta)F(\tau, \sigma)U(\Delta)^\dagger = F(\tau, \sigma + \Delta)$, where $F = x^i, X^I$, or S^a [one must recall that x^I and P^J satisfy (20)]. Thus, $U(\Delta)$ must equal unity when acting on physical states, on which (23) must then be satisfied.

The only truly new feature that enters into this story is the compactification and quantization of the extra 16 left-moving bosonic coordinates. It is this compactification, on a uniquely determined 16-dimensional compact space, that leads to the emergence of Yang–Mills interactions.

The extra 16 left-moving coordinates of the heterotic string can be viewed as parametrizing an "internal" compact space T. (This interpretation should not be taken too literally; in fact one can equally well represent these degrees of freedom by 32 real fermions.) The question then arises as to the nature of the internal manifold T. This is a dynamical question since a string theory is a theory of gravity. The choice of a background space-time is a dynamical issue—the background must be a solution of the string equation of motion. So far we have always imagined that the string lives in a flat, Minkowski, space-time. One might instead try to construct the theory about some other space-time manifold. In fact, in order to describe the real, four-dimensional world, we will surely have to do so. If one tries to embed the string in a general space-time background one finds inconsistencies. Thus the requirement of consistency of the first quantized theory is equivalent to demanding a solution of the equations of motion of the string theory. It was this procedure that fixed the dimension of space-time to be ten (for the superstring, 26 for the bosonic string) and, in the same fashion, will pick out the allowed backgrounds. It would be nice if the demand of consistency were to pick out a unique solution. This is certainly not the case, to date, for the ordinary space-time coordinates. The remarkable feature of the heterotic string is that the internal coordinates are required to lie in a space T that is completely determined to be a very special 16-dimensional torus (the maximal torus of $E_8 \times E_8$ or spin $32/Z_2$).

That a torus is a solution is reasonable since a torus is simply a flat space with periodic boundary conditions; however, there are many 16-dimensional tori. A generic 16-dimensional torus is characterized by 136 parameters (the radii of the 16 circles, whose product is the torus, and the angles between them). What picks out the special torus is the requirement that the coordinates of T are left moving (i.e., functions of $\tau + \sigma$). Consider the expansion of the internal coordinates $X^I(\tau + \sigma) = X^I + P^I\tau + L^I\sigma +$ oscillators, where X^I labels the position of the center of mass, and P^I is the total momentum. The momentum P^I is quantized (since X^I lives on a compact domain) in units of $1/R$ (where R is a radius of T).

The above expansion also contains a term linear in σ, which must be there if X^I is to be a function of $\sigma + \tau$. However, X^I is a periodic function of σ, so how can it have a term linear in σ? It must be that the coordinate comes back to the same point as we circle the string and σ goes from 0 to π. This is possible since the string lies on a torus and if the term $L^I\sigma$ goes around the torus (in some direction) an integer number of times, as σ goes from 0 to π, then X^I will be a periodic function of σ. Therefore L^I must equal an integer multiple of a radius of T in some direction. A string configuration with nonvanishing L^I represents a soliton, i.e., a string that winds around the torus some number of times, and its winding number is a topological, conserved charge. Such solitons are a new feature of string theories that do not occur in point theories. Now, in the heterotic string, X^I is a function of $\tau + \sigma$, so that L^I must equal P^I. This clearly restricts the form of the torus. Since $L^I \approx R$ and $P^I \approx 1/R$ it means that $R \approx 1$ (in our units this means $R \approx 1/M_{\text{Planck}}$). But the form of T is further constrained, since for a general torus it is not possible to identify the winding numbers, which span a lattice Γ, ($T = R_{16}/\Gamma$), with the momenta, P^I, which lie on the lattice Γ^* dual to Γ.

The lattice Γ is a set of vectors $[e_i^I; i = 1, \ldots, 16]$ which define the torus by the statement that points in R_{16} are identified according to

$$x^I \equiv x^I + \pi \sum_{i=1}^{16} e_i^I n_i \tag{24}$$

The momenta lie on the dual lattice, generated by e_i^{*I}; $i = 1, \ldots, 16$; where $\sum_I e_i^{*I} e_j^I = \delta_{i,j}$. This is in order to ensure that the translation operator, $e^{i2P\cdot\pi e_i}$, equals unity. Now, let us consider the constraints placed on the lattice, and thereby on the form of the torus T. First, the constraint (23) ensures that the square of any momentum $(P^I)^2$ be an even integer. Since $P^I = L^I$, this means that the lattice Γ is an even integer lattice, i.e., the square of the lattice vector is an even integer. Furthermore, every possible winding number L^I must correspond to an allowed momentum P^I. Thus $\Gamma \subseteq \Gamma^*$. Actually this last requirement is an automatic consequence of the fact that Γ is an even integer lattice.

It might seem obvious that the lattice and its dual must be identified in the case of the heterotic string. This is indeed so; however, one must examine radiative corrections to come to that conclusion. In fact the full consistency of the heterotic string theory requires that T be such that $\Gamma = \Gamma^*$, i.e., that the lattice defining the torus by *self dual.* If this is not satisfied then "modular invariance" breaks down. This means that the amplitudes, at the one loop level, develop anomalies that destroy reparametrization invariance.

Now, there exist very few self-dual lattices of the appropriate type; in fact they only exist in $8N$ dimensions! (The familiar hypercubic lattice with length equal to one is self dual, but not even.) In eight dimensions there is only one self-dual lattice—Γ_8, the root lattice of E_8. This lattice is a remarkable structure. It gives the most tightly packed arrangement of spheres in eight dimensions. Not surprisingly it is related to another beautiful structure; it is the root lattice of E_8, the most exceptional of all Lie groups. Cartan classified all Lie groups, and found that most lie in simple families [$SU(N)$, $SO(N)$, and $Sp(2N)$], but there were a finite number of exceptional groups (F_2, G_2, E_6, E_7, and E_8). Although hard to describe as a transformation group, E_8 is a beautiful mathematical object with most remarkable properties. The root lattice of this group is the lattice generated by the eight-dimensional vectors whose components are the eigenvalues of the eight set of commuting generators of E_8. In the case of SU_3 the root lattice is the familiar two-dimensional hexagonal lattice, from which one can read the allowed values of isospin and hypercharge. In eight dimensions the root lattice of E_8 is defined by the unimodular metric

$$g = \begin{pmatrix} 2 & -1 & 0 & 0 & 0 & 0 & 0 & 0 \\ -1 & 2 & -1 & 0 & 0 & 0 & 0 & 0 \\ 0 & -1 & 2 & -1 & 0 & 0 & 0 & -1 \\ 0 & 0 & -1 & 2 & -1 & 0 & 0 & 0 \\ 0 & 0 & 0 & -1 & 2 & -1 & 0 & 0 \\ 0 & 0 & 0 & 0 & -1 & 2 & -1 & 0 \\ 0 & 0 & 0 & 0 & 0 & -1 & 2 & 0 \\ 0 & 0 & -1 & 0 & 0 & 0 & 0 & 2 \end{pmatrix} \tag{25}$$

In 16 dimensions there exist two even self-dual lattices, $\Gamma_8 \times \Gamma_8$ and Γ_{16}, where Γ_{16} is the weight lattice of spin $32/Z_2$. Once we see that the torus on which the heterotic string lies is the maximal torus of these groups it is not surprising that they emerge as symmetries of the theory. To see this let us now examine the massless particles of the heterotic string. Since the theory contains gravity and since it possesses one supersymmetry (the one that transforms the supersymmetric right movers) it will contain the massless multiplet of ten-dimensional, $N = 1$, supergravity. We would also expect, in analogy to the standard Kaluza–Klein mechanism, that a compactified string theory will contain massless vector mesons associated with the isometries of the compact space. In Kaluza's original model one circle gave electromagnetism; here we have 16 circles and so we would expect 16 copies of electromagnetism. Indeed, since T has an isometry that consists of 16 $U(1)$'s, this compactification should, and does, yield the 16 gauge bosons of $U(1)^{16}$. The 16 conserved charges are just the 16 components of the total momentum, P^I, in the internal dimensions. These will be identified with electric charge, weak hypercharge, etc.

If this was all there was, a rather boring world would emerge. However, a remarkable feature of closed string theories is that for special choices of the compact space there will exist extra massless gauge bosons. These arise as massless solitons, namely, string configurations that wind around the internal torus, whose conserved charges are topological in nature. In the heterotic string there is an identification between the topological charges, the winding numbers L^I, and the nonvanishing momenta = winding number = $U(1)^{16}$ charge. If we look at the mass formula, (21), we see that there will exist such a particle for each vector on the lattice with length$^2 = 2$. There is precisely such a vector for each of the 480 charged generators of the algebra of $E_8 \times E_8$ or SO_{32}. These combine with the Kaluza–Klein gauge bosons to fill out the adjoint representation of a simple group whose rank equals the dimension of T. For the two allowed choices of T these produce the gauge bosons of $G = \text{spin}(32)/Z_2$ or $E_8 \times E_8$.

Once we have the gauge bosons it is not difficult to show that the full gauge symmetry is present. Thus we have seen that the heterotic string automatically produces, in ten dimensions, unique gauge interactions. The heterotic string theory has, by now, been developed to the same stage as other superstring theories. Interactions have been introduced, and shown to preserve the symmetries and consistency of the theory, and radiative corrections have been calculated and shown to be finite. In many ways it now appears as the simplest of all superstring theories. It surely provides a most satisfactory explanation for the emergence of specific gauge interactions and, as we have seen, offers much phenomenological promise.

DISCUSSION

N. BRALIĆ: The subtraction of minus one in the expression for $\tilde{N}$ seemed like fine tuning, arranged just to cancel the tachyon. Is the removal of the tachyon state in the heterotic string natural?

D. GROSS: Actually, the subtraction of minus one is what gives rise to the tachyon in the case of the bosonic string. The removal of the tachyon occurs because of the constraint, (23), that equates the left- and right-moving number operators. I discussed the geometrical meaning of this constraint above. States that do not satisfy this constraint, including the would-be tachyon, are simply not reparametrization invariant.

J. ALFARO: Is the issue of anomalies in string theories related to ultraviolet behavior, as it is in ordinary field theory?

GROSS: That is an interesting issue which is not totally understood. String theories of the closed variety, including heterotic strings, simply have no ultraviolet divergences. It appears that if one attempts to write down an anomalous heterotic string theory one finds violations of unitarity, as if regulator negative metric fields were present. However, these are not required to render the theory finite.

3. STRING EQUATIONS OF MOTION

String theories are typically formulated in flat space-time. However, if they are to have any chance of describing the real world then we must find other solutions, other consistent quantum vacuum states that look more like our real world, in which six of the spatial dimensions are curled up into a small compact manifold. Since string theory is a theory of gravity this is a dynamical question. The nature of the space-time manifold is a dynamical issue. We must find a space-time manifold in which the string propagates, which is itself a true solution of the gravitational or string equations of motion. In this lecture I shall discuss three approaches to deriving the equations of motion that can be used to derive the background space-time and the relation between them.

What we are looking for is a solution of the classical equations of string theory. This is not because we are interested in classical string theory as such, but rather that such a solution is the starting point for the construction of a consistent quantum ground state. But first we have to construct the classical equations of motion. There are three separate approaches to this problem. These are based on string field theory, on the effective Lagrangian derived from the S matrix, and on the conformal invariance of the σ models that describe first quantized strings.

The first approach, which is perhaps the most correct one, is based on string field theory. Unfortunately, this theory is only in its embryonic stage of development. In particular, for closed strings an adequate string Hamiltonian only exists in light cone gauge. In such a formulation of string theory one would simply look for extrema of the action that defines the theory. Let us examine what these equations look like in light cone gauge, where we already possess the Hamiltonian.

In light cone gauge the string is defined by a *functional* $\Psi[x^i(\sigma), x^+]$ which depends on the physical degrees of freedom of the string—its transverse spatial coordinates, and on light cone time x^+. In a very schematic form the Hamiltonian is

$$H = \int \mathscr{D}x^i(\sigma)\, dx^+ \Psi\left[\frac{\delta^2}{\sigma x^2} + x'^2\right]\Psi + \iiint \Psi(1)\Psi(2)\Psi(3)\delta\{x(1) - [x(2) \oplus x(3)]\} \tag{26}$$

The interaction represents the joining of two strings (2 and 3) at a point to form a third string (1). This Hamiltonian does yield the Feynman rules of the second quantized theory, from which can calculate all the scattering amplitudes of the string states. Similar expressions can be written for open string field theories, as well as for superstrings and for the heterotic string.

In principle one could use this light cone gauge action to derive the string equations of motion. Of course light cone gauge is rather nasty if one is looking for solutions that correspond to curved manifolds. However, if one were exploring vacuum solutions, where the background might be four-dimensional Minkowski space times an internal curved manifold, then there is no problem in singling out the coordinates $x^0 \pm x^3$, and this formalism could, in principle, be used to search for solutions. The real problem is that the equations are unfamiliar and difficult. They are functional integral-differential equations, with which we have little experience. Even though we know the equations and we know many solutions, nobody has managed to recover the known solutions from the known equations. The covariant form of these equations might be more useful. It does not hide the symmetries of the theory and is certainly more elegant. However, even in the framework of the covariant approach little has been done with the formalism except to show that one can recover perturbation theory. So, whereas string field theory is probably required for a nontrial understanding of string dynamics, to date most information has been garnered by other approaches.

The difficulty with string field theory is that the string field contains many, many particles. Even so, it can, in principle, be decomposed into its pointlike constituents. One can replace a string theory by a theory of an infinite set of particles whose couplings and masses are all related in a definite way. The advantage of focusing on the point field dynamics of the particles is that we are much more familiar with field theories that have been explored by physicists for centuries than with functional field theories.

One could start with the functional field of the string $\Psi[x(\sigma)]$ and decompose it into local fields,

$$\Psi[x(\sigma)] = \sum_{n_1 n_2 \ldots} \Psi_{n_1 n_2 \ldots}(x_{\text{com}}) \Phi_{n_1 n_2 \ldots}(x_n) \tag{27}$$

where $\Phi_{n_1 n_2 \ldots}(x_n)$ span a complete basis for the string functionals (say the Fock space wave functions of the free string), and $\Psi_{n_1 n_2 \ldots}(x_{\text{com}})$ are local fields (one for each string particle) that are *functions* of the center-of-mass coordinate x_{com}. Plugging this decomposition into the string Lagrangian, or into its covariant version, would yield a local action for the infinite number of local fields that represent the full string. The resulting action would only have cubic interactions, at the price, however, of involving an infinite number of local fields. One might wonder how a theory with only cubic vertices can reproduce the interactions of gravitons, which according to Einstein's theory have vertices of arbitrarily high order. The answer is that the quartic and higher-order graviton couplings emerge as a low-energy approximation to the exchange of the massive string modes, much as the

Fermi theory of weak interactions emerges as a low-energy approximation to the exchange of massive vector bosons.

Although cubic, this formalism is not very convenient since we are simply not used to dealing with an infinite number of fields. Now, some of the fields are massless (the graviton, the gauge mesons, etc.), but most correspond to massive degrees of freedom of the string, which cannot be seen directly at low energies. It is therefore convenient, at least for describing the low-energy behavior of the theory, to eliminate all but the massless modes. In principle this could be done exactly, at the price of generating horrible nonpolynomial (and eventually nonlocal) interactions for the massless modes. In practice this procedure can be carried out perturbatively in powers of the relevant momenta compared to the Planck mass. Imaging starting from the infinite set of cubic equations for the massless (m) fields and for the massive ($\mathcal{M}$) fields. These obey equations which are schematically

$$\begin{aligned} \Delta m &= m^2 + \mathcal{M}^2 + \mathcal{M}m \\ (\Delta + M_{\text{Planck}}^2)\mathcal{M} &= \mathcal{M}^2 + m^2 + \mathcal{M}m \end{aligned} \tag{28}$$

Now, we can systematically eliminate $\mathcal{M}$ by solving for it in terms of m

$$\mathcal{M} = \frac{1}{M_{\text{Planck}}^2} m^2 - \frac{\Delta}{M_{\text{Planck}}^4} m^2 + \frac{m^4}{M_{\text{Planck}}^4} + \cdots \tag{29}$$

in an expansion in powers of $\Delta / M_{\text{Planck}}^2$ and $m^2 / M_{\text{Planck}}^2$. Then one can plug these solutions back into the equations for m. This would give equations for the massless fields which would arise from an effective Lagrangian with arbitrarily high powers of the fields and of their derivatives. Note that one is not setting the massive fields equal to zero; rather one is solving for them in terms of the massless fields, i.e., integrating them out. Of course, one could always make a field redefinition of the massive fields, setting $\mathcal{M}' = \mathcal{M} - (1/M_{\text{Planck}}^2)m^2 - \ldots$, thereby setting $\mathcal{M}'$ to zero.

Actually, this procedure has never been fully carried out; rather, one has used knowledge of the S matrix of string theory to construct an effective Lagrangian for the component massless fields. Given the S matrix that describes the scattering of massless string states about, say, flat space one *guesses* (actually there is a systematic procedure for doing this guessing) a Lagrangian that reproduces the S matrix. This proceeds as follows. First, starting with the free spectrum we already know what the kinetic terms in the Lagrangian are. Then by examining the cubic vertices we can deduce the cubic couplings of the massless particles. From these pieces we can construct an effective Lagrangian to lowest order, $\mathcal{L}_0$. Next, we examine the scattering amplitudes for the massless particles. Part of these will already be consequence of $\mathcal{L}_0$, which generates various Feynman diagrams that contain massless poles. In fact, according to unitarity all the massless poles are generated by $\mathscr{L}_0$. But there is more in the amplitudes than just these massless poles; after all, there are all the infinite number of massive states, which are not described by $\mathscr{L}_0$. However, these states, being heavy, give contributions that are regular (as functions of the external momenta) for small momenta. Therefore, the left-over pieces, not accounted for by $\mathscr{L}_0$, can be expanded in a power series in p^2/M^2_{Planck}. The terms in this expansion correspond to terms in an effective Lagrangian, just as the exchange of a heavy vector meson between an electron and a neutrino can be expanded in powers of the momentum transfer, thus yielding an infinite series of four fermion interactions, beginning with the Fermi interaction.

This procedure is quite simple and has been exploited to yield the effective Lagrangian for superstring theory and for the heterotic string theory up to quartic order. One of the reasons it is simple is that one can exploit the symmetries of the theory to figure out many higher-order terms in the effective Lagrangian once lower-order terms are known. Thus, once we know that the theory contains massless spin-two particles, gravitons, then it must contain, at the very least, all of the vertices that follow from the Poincaré–Einstein Lagrangian, and once you know that you have massless charged vector bosons then you must have the full Yang–Mills Lagrangian (plus perhaps higher-order interactions).

It might be objected that the Lagrangian derived in this fashion is not unique—after all there may be many Lagrangians that give the same S matrix. True, there are many equally good Lagrangians. In fact, by simply redefining the fields one can change the effective Lagrangian. Imagine that we have a bunch of particles and we are given the S matrix that describes their scattering. Now associate with each a local field ϕ^i and construct a Lagrangian that reproduces the S matrix, $\mathscr{L}(\phi^i)$. The fields ϕ^i are just coordinates of the configuration space of this quantum theory and we are certainly allowed to make a change of coordinates without modifying the physics. If we change ϕ^i to $\phi^{i\prime}$, by means of a nonsingular point transformation: $\phi^i \to \phi^{i\prime}(\phi^i)$, then the form of the Lagrangian changes

$$\mathscr{L}[\phi^i] = \mathscr{L}[\phi^i(\phi^{i\prime})] \equiv \mathscr{L}'[\phi^{i\prime}] \tag{30}$$

but the S matrix is unchanged. So the effective Lagrangian is clearly highly nonunique. But this nonuniqueness is irrelevant—it does not affect physics. What is physics? Well, physics is the S matrix, for example, which is clearly unchanged. But physics is also more. We are interested not just in calculating the S matrix about flat space, from which we guessed the Lagrangian, but also in the classical equations of motion that follow from the Lagrangian. These can be used to search for new solutions, which would be starting points for the construction of new quantum ground states. Although the form of the Lagrangian changes, when we redefine the fields, the solution set of the equations of motion remains the same since

$$\frac{\delta \mathscr{L}}{\delta \phi_i} = \frac{\delta \mathscr{L}'}{\delta \phi_j'} \frac{\delta \phi_j'}{\delta \phi_i} \tag{31}$$

so that as long as the change of variables is nonsingular the zeroes of $\delta\mathscr{L}/\delta\phi_i$ coincide with those of $\delta\mathscr{L}'/\delta\phi_i'$.

Finally, the third method of constructing the string theory equations of motion is indirect. It is based on the first quantization of a string, which is now taken to be embedded in some nontrivial space-time background. The action that governs the dynamics of such a string is

$$S = \int d^2\xi \sqrt{g} g^{\alpha\beta} \, \partial_\alpha x^\mu \, \partial_\beta x^\nu G_{\mu\nu}(x^\mu) \tag{32}$$

where $G_{\mu\nu}$ is the metric of the manifold in which the string is embedded. If we set $G_{\mu\nu} = \eta_{\mu\nu}$, where $\eta_{\mu\nu}$ is the Minkowski metric, this is simply the Nambu action and its quantization leads to the usual bosonic string. Actually, the consistency of that string theory, in 26 dimensions, is telling us that Minkowski space is a solution of the string theory equations of motion. For an arbitrary $G_{\mu\nu}$ one will not have a consistent first quantized string, and satisfying the general requirements of consistency will be equivalent to satisfying the string equations of motion. The above Lagrangian, for a generic $G_{\mu\nu}(x)$, is much more difficult to deal with since it describes a nonlinear, interacting σ model. This is a nontrivial quantum field theory, complete with ultraviolet divergences and renormalization.

As a renormalizable field theory the Lagrangian of (32) is clearly not sufficient. One must, according to the usual rules of renormalization theory, include in the Lagrangian all possible operators of dimension two (we are working in two dimensions), since they will in any case be generated by enormalization. Thus, in addition to the metric term above we must add (for the bosonic string) the terms

$$\delta S = \int d^2\xi [\varepsilon^{\alpha\beta} \, \partial_\alpha x^\mu \, \partial_\beta x^\nu B_{\mu\nu} + \sqrt{g} R^{(2)} \Phi + \cdots] \tag{33}$$

where $B_{\mu\nu}$ is the antisymmetric (torsion) field, $R^{(2)}$ is the two-dimensional curvature, and Φ is the dilation field. The heterotic string theory also can have additional terms corresponding to background gauge fields. For each massless mode of the string theory there is a corresponding term in the Lagrangian. Massive background modes would correspond to nonrenormalizable terms in the σ model Lagrangian, and nobody, to date, knows how to handle these (for the solutions that have been found so far they are not necessary).

What are the consistency conditions that determine $G_{\mu\nu}$? They are the conditions that the string be invariant under conformal transformations. Remember that it was the two-dimensional conformal invariance of the string that led to the gauge symmetries of the resulting field theories and decoupled the unphysical states. What does this requirement mean? Even for the string in flat space it was a nontrivial requirement. In fact, it was the condition of conformal invariance that led to the critical dimension of 26 (for the bosonic string) or 10 (for the superstring). In the case of a curved background, where the quantum mechanics of the string is described by a renormalizable quantum field theory, conformal invariance is likely to break down owing to renormalization effects. In order to have a conformally invariant theory we require that all the *β functions* vanish, i.e., that the theory is *finite.* The β functions, it will be recalled, tell us how the dimensionless couplings, which should be scale invariant, change with a change in scale. Even the condition that flat space be 26 (or 10) dimensional can be regarded as following from the requirement that the β function of the dilaton vanish.

Let us indicate how one goes about calculating these β functions and how Einstein's equations emerge. Consider the purely bosonic theory of the closed string described by (32). This is a nonlinear σ model in which the two-dimensional coordinates ξ^α are mapped into a curved *target space* $x^\mu \in \mathcal{M}$ with metric $G_{\mu\nu}$. In this theory the function $G_{\mu\nu}(x^\mu)$ plays the role of the coupling constants. In other words, if we expand $G_{\mu\nu}(x^\mu)$ in a power series in x^μ, about some arbitrary point, we would have an infinite number of nonlinear interactions. It is more useful to think about the whole function $G_{\mu\nu}(x^\mu)$ as a local coupling constant and to calculate $\beta_{G_{\mu\nu}(x^\mu)}$. Because of locality, this will be a local function of $G_{\mu\nu}(x^\mu)$ and its derivatives.

Let us calculate the one-loop contribution to this β function. The above theory is invariant under coordinate changes of x^μ (with $G_{\mu\nu}$ transforming in the usual way), since this is just a redefinition of the field $x^\mu(\xi)$. We can exploit this freedom to choose to expand in *normal coordinates* about the point $x_0^\mu = 0$, whereby

$$G_{\mu\nu}(x) = G_{\mu\nu}(0) + R_{\mu AB\nu}x^A x^B + \cdots \tag{34}$$

If we plug this into (32) the first term defines the propagator and the second yields a quartic interaction,

$$S = \int [G_{\mu\nu}(0)\,\partial_\alpha x^\mu\,\partial_\alpha x^\nu + R_{\mu AB\nu} + x^A x^B\,\partial_\alpha x^\mu\,\partial_\alpha x^\nu + \cdots] \tag{35}$$

Let us now calculate the one-particle irreducible two-point function, which starts out as the inverse bare propagator, $G_{\mu\nu}(x_0)$. The one-loop correction is proportional to

$$\int^{\Lambda} R_{\mu\alpha\beta\nu}(x_0)G^{\alpha\beta}/p^2 d^2p \sim R_{\mu\nu}(x_0)\ln(\Lambda) \tag{36}$$

where Λ is an ultraviolet cutoff. To cancel this divergence we must add a counterterm to $G_{\mu\nu}(x_0)$ of the form $R_{\mu\nu}(x_0)\ln\Lambda$. Now recall that $\beta_g = \Lambda\partial_\Lambda g$. It then follows that

$$\beta_{G_{\mu\nu}(x)} = R_{\mu\nu}(x) + \cdots \tag{37}$$

We find that the β function to lowest order equals the Ricci tensor. If we wish to embed the string in a curved space then we must choose it to be Ricci flat in order to preserve conformal invariance. In this way one finds that the string equations of motion are, to lowest order, identical to Einstein's. If we allowed for the possibility that the fields $B_{\mu\nu}$, $\Phi, \ldots$ had nonvanishing expectation values then there would be other terms in β_G arising from the interactions contained in (33). In addition, demanding that $\beta_{B_{\mu\nu}} = 0$, $\beta_\Phi = 0, \ldots$ would produce equations of motion for $B_{\mu\nu}$, $\Phi, \ldots$.

In this fashion we can determine the string equations of motion, by calculating the β functions of the relevant σ model. This yields an expansion of the equations of motion in perturbation theory. What are we perturbing in? The expansion parameter is not the loop (or quantum) expansion parameter of the string theory. Rather it is the curvature of the target manifold, or its inverse radius (R), in units of the Planck length. This is the same expansion parameter that we had before since a typical momentum is of order $1/R$. Now, in the case of the bosonic theory, it would appear very difficult to find solutions of these equations to all orders. To lowest order the manifold must be Ricci flat and there are many such solutions. But at two-loop order this no longer suffices since the β function receives contributions proportional to the square of the Riemann tensor. In fact, if we demand that the one and two-loop contributions vanish separately then we find that the manifold must be flat.

It is conceivable that a very special manifold does yield a conformally invariant σ model, as a consequence of cancellations between different orders of perturbation theory. This is unlikely. Much easier to find are theories that have a *line of critical points*, i.e., a continuous family of conformally invariant theories. In particular if this family goes over continuously to free field theory then we could justify our use of perturbation theory. This is the case if the β functions vanish order by order in perturbation theory. In the case of the heterotic string the corresponding σ model is conformally invariant, order by order in perturbation theory, if the (six-dimensional internal) manifold is a Calabi–Yao manifold, i.e., Ricci flat and of $SU(3)$ holonomy. (Actually this statement must be slightly modified; see below.)

In the σ-model approach we do not derive a Lagrangian for string theory, only the equations of motion. Furthermore, these are highly non-unique since the β functions of a quantum field theory are not unique. If one redefines the couplings of a theory then the β functions change. In our case the couplings are identified with the local massless fields $(G_{\mu\nu}, \Phi, \ldots)$, which we denote again collectively as ϕ^i. Under a change $\phi \to \phi(\phi')$ the β functions, $\beta_i(\phi) \equiv \Lambda(\partial/\partial\Lambda)\phi_i$, vary according to

$$\beta_i(\phi) \equiv \Lambda \frac{\partial}{\partial \Lambda} \phi_i(\phi') = \beta_j'(\phi') \frac{\delta \phi_i}{\delta \phi_j'} \tag{38}$$

However, the zeroes of β_i, here identified with the equations of motion, are invariant under such a redefinition. It is clear that this ambiguity is the same as the ambiguity discussed above in the derivation of the effective Lagrangian. It is also clear that these equations, $\beta_i = 0$ and $\delta\mathcal{L}/\delta\phi_i = 0$, must be identical. Otherwise, since both must be satisfied, the fields would be overconstrained. This suggests the β functions are related to the effective Lagrangian by

$$\beta^i(\phi) = \Gamma^{ij} \frac{\delta \mathcal{L}}{\delta \phi_j} \tag{39}$$

where Γ^{ij} is similar to a *metric* in field space, and transforms under field redefinitions according to

$$\Gamma^{lk}(\phi) = \Gamma^{ij}(\phi') \frac{\delta \phi_l'}{\delta \phi_i} \frac{\delta \phi_k'}{\delta \phi_j}$$

A direct proof of this equivalence is lacking, although it is certainly true to low orders where explicit comparisons have been made.

During the last year much effort has been devoted to exploring the classical equations of motion of string theory. Mostly one wishes to find static solutions that might describe the background fields of stable quantum vacua. Both of the two approaches to the derivation of these equations were explored. The first method, that of deducing the effective action from the string scattering amplitudes, has been pursued by Witten and myself; the second, based on demanding conformal invariance of the two-dimensional σ model that describes the first quantized string in a nontrivial background space, has been discussed by Grisaru, Zanon, deVen, Pope, and Freedman. Both groups agree that, to quartic order in the curvature (say of the internal manifold), the effective Lagrangian (in the type II superstring or in the heterotic string with the spin connection embedded in the gauge group) differs from the Einstein Lagrangian, $\sqrt{g}R$, by terms quartic in the Riemann tensor R_{abcd}. These terms have an interesting effect on the issue as to whether the Calabi–Yau spaces are indeed solutions of the string equations to all orders. Strictly speaking, they invalidate previously given "proofs" to that effect. However, one can show that for every Calabi–Yau space there is another space close by (i.e., one whose metric differs by terms that are small, when the curvature of the internal manifold is small, and that can be calculated order by order in an expansion in the curvature). This requirement, to slightly shift the metric of the solution, has no effect on the resulting physics (the number of generations, the breaking or preservation of symmetries, etc.).

Other attempts have been made, notably by Witten, Strominger, and Hull, to extend the class of solutions to include ones where the spin connection is not embedded in the gauge group. This requires turning on torsion. It has the advantage of allowing a GUT group of $SU(5)$ or $SO(10)$, which solves many phenomenological problems. Unfortunately, almost all such solutions have (σ model) nonperturbative corrections to the β function (generated by σ model instantons), which destroy conformal invariance.

DISCUSSION

L. BAULIEU: If you keep going and renormalize the σ model to arbitrary high order won't you generate an infinite number of counterterms and in effect produce nonlocal counterterms?

D. GROSS: That is correct, and corresponds to the nonlocality that must arise in string theory when the momenta are of order the Planck mass and we see the nonlocal effects of massive string state exchanges.

ACKNOWLEDGMENT. This work was supported in part by the National Science Foundation under grant No. PHY 80/19754.

Chapter 10

BRST Symmetry in the Classical and Quantum Theories of Gauge Systems

Marc Henneaux

1. INTRODUCTION

In the last few years, it has become increasingly clear that Becchi-Rouet-Stora-Tyutin methods [1, 2] play a fundamental role in modern gauge field theory. These methods are not only useful in the path integral formalism, where the BRST generator is an essential building block of the effective action, but also they find important applications in the operator formulation of the theory. For instance, in string theory, one finds that nilpotency of the BRST quantum generator only holds in 26 space-time dimensions for the bosonic model [3] and in 10 space-time dimensions for the fermionic one [4]. This provides an alternative and very economical way for deriving the string critical dimensions.

The purpose of this chapter is to discuss the Hilbert space aspects of the BRST symmetry, as well as its classical roots. Because Hamiltonian methods are close to the operator formulation of quantum mechanics, we will rely on them throughout.

The main results reported here are due to Fradkin and his collaborators in a remarkable series of papers [5-8] (general construction of the BRST

MARC HENNEAUX • Faculté des Sciences, Université Libre de Bruxelles, B-1050, Bruxelles, Belgium; and Centro de Estudios Científicos de Santiago, Santiago 9, Chile.

generator), to Kugo and Ojima [9], and to Curci and Ferrari [10] (operator formalism). These works have been reviewed in Ref. 11. The author's contributions in the field appear to be (i) explicit demonstrations of theorems on the existence of the BRST generator and of BRST invariant extensions of gauge invariant observables [11, 12]; (ii) classical interpretation of the ghosts as making manifest the canonical covariance of the structure of gauge systems [11], and (iii) demonstration of the equivalence between the physical subspaces of the BRST and Dirac approaches in simple cases [11; see also 13].

The chapter is organized as follows: First, the classical properties of the BRST transformation are studied. A clear distinction between the classical and quantum aspects of the BRST symmetry is indeed necessary if one wants to distinguish between purely quantal effects (anomalies in the BRST algebra) and general features that exactly hold classically.

The structure of constrained Hamiltonian systems, within which scope gauge theories fall [14], is studied. Structure functions are defined and shown to obey remarkable identities which imply the existence of the classical BRST generator Ω. This generator, which involves the ghosts, is real and nilpotent ($[\Omega, \Omega] = 0$). It exists for theories with a closed or open algebra, and furthermore, it has an intrinsic meaning, i.e., *no gauge condition is necessary to construct* Ω.

The existence of BRST invariant extensions for any gauge invariant observable is then demonstrated. The ambiguity in the structure functions is discussed and is completely accounted for by a canonical transformation in the extended phase space. BRST invariant extensions of gauge invariant observables are also shown to be determined up to the addition of a term like $[K, \Omega]$, where K is an arbitrary function of ghost number -1. The analysis of the classical theory ends with a comparison between classical trajectories in the original and extended phase spaces.

The second part of the chapter is devoted to the quantum theory. The subsidiary condition that defines the physical subspace is introduced and motivated by the demand that the above-mentioned ambiguity in the BRST invariant observables be physically irrelevant.

It is found that a new kind of gauge invariance, called "quantum gauge invariance," appears, provided that there is no anomaly in the BRST algebra. The quantum gauge invariance is closely related to (but is bigger than) the original gauge invariance. As a result of this new invariance, the number of physical modes in the quantum and classical theories can be shown to be equal for simple models.

For more complicated systems, it is pointed out that a "doubling" of BRST states can occur [15, 3]. The physical implications of this phenomenon are discussed. The Appendix discusses the "doubling" phenomenon in the case of the string models.

2. STRUCTURE OF CONSTRAINED HAMILTONIAN SYSTEMS

2.1. Zeroth-Order Structure Functions

Because of the gauge invariance of the action, the Legendre transformation between the velocities and the canonically conjugate momenta is singular in the case of gauge theories. How to handle this problem *without fixing the gauge* has been shown by Dirac [14].

The main feature of the Dirac formalism is the appearance of constraints between the coordinates and the momenta,

$$G_a(q, p) = 0, \qquad a = 1, \ldots, m \tag{1}$$

which we assume for notational simplicity to be all commuting. Here, (q, p) stands for (q^i, p_i), $i = 1, \ldots, n > m$.

Let us stress that the constraints (1) solely follow from the gauge invariance of the action and have nothing to do with gauge fixing conditions. They reflect a fundamental feature of gauge theories. Examples of constraints like (1) are (i) Gauss' law in electromagnetism and Yang–Mills theory; (ii) "super-Hamiltonian" and "supermomentum" constraints of general relativity, which lead, in the quantum theory, to the famous Wheeler–De Witt equations; (iii) the Virasoro conditions in string theory; (iv) the "mass-shell" condition $p^2 + m^2 = 0$ for the relativistic particle, etc.

The constraints will be called in the sequel "zeroth-order structure functions," and sometimes denoted by $\overset{(0)}{U}_a(q, p)$ because they are only the first ingredient of an entire series of interesting structure functions.

The purpose of the constraints (1) is twofold. On the one hand, they restrict the classically available portion of phase space by the equations $G_a(q, p) = 0$. On the other hand, they are the canonical generators of the gauge transformations.

The zeroth-order structure functions are actually not completely determined by the action itself and involve some ambiguity. What is well defined is the constraint surface $G_a = 0$, which can be described by different equivalent set of equations. Two such sets are related by

$$\overset{(0)}{U'}_a(q, p) = a_a^b(q, p)\, \overset{(0)}{U}_b(q, p), \qquad \det a_a^b \neq 0 \tag{2}$$

The choice of the new functions $\overset{(0)}{U'}_a(q, p)$ as zeroth-order structure functions also amounts to a change in the description of the gauge transformations.

It is clear that the brackets $[\overset{(0)}{U'}_a, \overset{(0)}{U'}_b]$ are in general different from $[\overset{(0)}{U}_a, \overset{(0)}{U}_b]$. So, one cannot go from $\overset{(0)}{U}_a$ to $\overset{(0)}{U'}_a$ by a mere canonical transformation in the original phase space (q^i, p_i). It will be one virtue of the ghosts to yield a canonical interpretation of (2).

2.2. First-Order Structure Functions

If one applies a gauge transformation to a point on the constraint surface, one must remain on the constraint surface. Otherwise it would not be permissible to perform gauge transformations. This would contradict the assumption of gauge invariance.

The constraints are accordingly "first class,"

$$[\overset{(0)}{U}_a, \overset{(0)}{U}_b] = -2\overset{(1)}{U}{}^c_{ab}\overset{(0)}{U}_c \tag{3}$$

The functions $\overset{(1)}{U}{}^c_{ab}(q, p) = -\overset{(1)}{U}{}^c_{ba}(q, p)$, defined by (3), are the "first-order structure functions" (the factor of -2 has been inserted for further convenience).

The "first-order structure functions" also carry some ambiguity, since (3) determines them only up to

$$\overset{(1)}{U}{}^c_{ab} \to \overset{(1)}{U}{}'^c_{ab} = \overset{(1)}{U}{}^c_{ab} + M^{cd}_{ab}\overset{(0)}{U}_d \tag{4a}$$

$$M^{cd}_{ab} = -M^{dc}_{ab} = -M^{cd}_{ba} \tag{4b}$$

Furthermore, the $\overset{(1)}{U}{}^c_{ab}$'s transform according to some complicated rule not needed here under (2).

When the functions $\overset{(1)}{U}{}^c_{ab}$ turn out to be constant, as in the Yang–Mills theory, one has a true gauge group with a closed algebra. When they depend on the canonical variables, one says that the gauge "algebra" is "open," for the commutator of two gauge transformations generated by the constraints is a transformation of the same type only on the constraint surface (the commutator involves the term $[\ldots, \overset{(1)}{U}{}^c_{ab}]\overset{(0)}{U}_c$, which only vanishes weakly).

2.3. Second-Order Structure Functions

From the Jacobi identity for the Poisson brackets, it follows that

$$\sum[[\overset{(0)}{U}_a, \overset{(0)}{U}_b], \overset{(0)}{U}_c] = 0 \tag{5}$$

where one sums over all cyclic permutations of (a, b, c). This implies, when (3) is taken into account, that

$$\overset{(0)}{U}_d\{\sum([\overset{(1)}{U}{}^d_{ab}, \overset{(0)}{U}_c] + 2\overset{(1)}{U}{}^e_{ab}\overset{(1)}{U}{}^d_{ce})\} = 0 \tag{6}$$

This relation, in turn, leads to [11]

$$[\overset{(1)}{U}{}^d_{[ab}, \overset{(0)}{U}_{c]}] + 2\overset{(1)}{U}{}^e_{[ab}U^{(1)}{}^d_{c]e} = 2\overset{(2)}{U}{}^{dm}_{abc}\overset{(0)}{U}_m \tag{7}$$

where the functions $\overset{(2)}{U}{}^{mm}_{abc}(q, p)$ are antisymmetric in (m, n) and (a, b, c) and are named "second-order structure functions." [The subscript brackets in (7) denote, as usual, complete antisymmetrization.]

Although the second-order structure functions are easily seen to vanish for Yang–Mills theories (as a result of the Jacobi identity for the structure constants) and gravity, they are nonzero in the generic case of an open algebra. Hence, they cannot be neglected.

Again, the second-order structure functions involve some ambiguity. They are determined by $\overset{(0)}{U}_a$ and $\overset{(1)}{U}{}^a_{bc}$ only up to

$$\overset{(2)}{U}{}^{mm}_{abc} \to \overset{(2)}{U}{}'^{mm}_{abc} = \overset{(2)}{U}{}^{mm}_{abc} + M^{mmp}_{abc}\overset{(0)}{U}_p \tag{8}$$

where $M^{mmp}_{abc}(q, p)$ is completely antisymmetric in both the upper and lower indices.

2.4. Higher-Order Structure Functions

It is remarkable that one can repeat the above construction of the second-order structure functions step by step and generate in this way all the "higher-order structure functions."

The procedure goes as follows in the case of the third-order structure functions: If one takes the Poisson bracket of the defining equation (7) with the zeroth-order structure functions $\overset{(0)}{U}_d$, completely antisymmetrizes the resulting expression with respect to (a, b, c, d), systematically uses the Jacobi identity for the Poisson brackets and the defining equations (7) and (3), one gets the identity

$$\overset{(2)}{D}{}^{[a_1a_2]}_{[b_1b_2b_3b_4]}\overset{(0)}{U}_{a_2} = 0 \tag{9}$$

with

$$\begin{aligned}\overset{(2)}{D}{}^{a_1a_2}_{b_1b_2b_3b_4} = & -\frac{1}{2}\sum_{q=0}^{2}\left[\overset{(q)}{U}{}^{a_1\ldots a_q}_{b_1\ldots b_{q+1}}, \overset{(2-q)}{U}{}^{a_{q+1}\ldots a_2}_{b_{q+2}\ldots b_4}\right] \\ & - \sum_{q=0}^{1}(q+1)(3-q)\overset{(q+1)}{U}{}^{a_1\ldots a_q c}_{b_1\ldots b_{q+2}}\overset{(2-q)}{U}{}^{a_{q+1}\ldots a_2}_{b_{q+3}\ldots b_4 c}\end{aligned} \tag{10}$$

From (9), it follows that [11]

$$\overset{(2)}{D}{}^{[a_1a_2]}_{[b_1b_2b_3b_4]} = 3\overset{(3)}{U}{}^{a_1a_2a_3}_{b_1\ldots b_4}\overset{(0)}{U}_{a_3} \tag{11}$$

where the "third-order structure functions" are completely antisymmetric in both the a and b indices.

For the higher-order structure functions, one finds, in an analogous manner, the identities

$$\overset{(n)}{D}{}^{[a_1 \dots a_n]}_{[b_1 \dots b_{n+2}]} \overset{(0)}{U}_{a_n} = 0 \tag{12}$$

which imply in turn the existence of the structure functions $\overset{(n+1)}{U}{}^{a_1 \dots a_n a_{n+1}}_{b_1 \dots b_{n+2}} \equiv \overset{(n+1)}{U}{}^{[a_1 \dots a_n a_{n+1}]}_{[b_1 \dots b_{n+1}]}$ or order $n+1$,

$$\overset{(n)}{D}{}^{[a_1 \dots a_n]}_{[b_1 \dots b_{n+2}]} = (n+1) \overset{(n+1)}{U}{}^{a_1 \dots a_n a_{n+1}}_{b_1 \dots b_{n+2}} \overset{(0)}{U}_{a_{n+1}} \tag{13}$$

In (12) and (13), $\overset{(n)}{D}{}^{a_1 \dots a_n}_{b_1 \dots b_{n+2}}$ stands for

$$\overset{(n)}{D}{}^{a_1 \dots a_n}_{b_1 \dots b_{n+2}} = \frac{1}{2} \sum_{q=0}^{n} [\overset{(q)}{U}{}^{a_1 \dots a_q}_{b_1 \dots b_{q+1}}, \overset{(n-q)}{U}{}^{a_{q+1} \dots a_n}_{b_{q+2} \dots b_{n+2}}](-)^{nq+1}$$
$$- \sum_{q=0}^{n-1} (q+1)(n-q+1) \overset{(q+1)}{U}{}^{a_1 \dots a_q c}_{b_1 \dots b_{q+2}} \overset{(n-q)}{U}{}^{a_{q+1} \dots a_n}_{b_{q+3} \dots b_{n+2} c} (-)^{n(q+1)} \tag{14}$$

and only involves structure functions of order $\leq n$.

The explicit demonstration of the identities (12), without which it would be impossible to infer (13) and hence to construct the higher-order structure functions, can be found in Refs. 11 and 12 (see also Ref. 16 in a related context). Let us simply note that for $n = 1$ and 2, the relations (12) and (13) reduce to (6), (7), and (3). The existence of the structure functions will be demonstrated in a more indirect way below.

Finally, one easily checks that (13) determines the structure functions of order $n+1$ only up to

$$\overset{(n+1)}{U}{}^{a_1 \dots a_{n+1}}_{b_1 \dots b_{n+2}} \to \overset{(n+1)}{U}{}^{a_1 \dots a_{n+1}}_{b_1 \dots b_{n+2}} + M^{a_1 \dots a_{n+1} a_{n+2}}_{b_1 \dots b_{n+2}} \overset{(0)}{U}_{a_{n+2}} \tag{15}$$

(for given structure functions of order $\leq n$), where $M^{a_1 \dots a_{n+2}}_{b_1 \dots b_{n+2}}$ possesses the appropriate antisymmetry properties.

3. THE CLASSICAL GENERATOR OF THE BRST TRANSFORMATION

3.1. BRST Generator as "Generating Function" of the Structure Functions

By introducing anticommuting degrees of freedom, one can summarize all the structure relations (13) and the identities (12) in the concise statement that there exists an anticommutating generator, the "BRST generator," which is nilpotent.

The procedure goes as follows. First, one associates with each constraint G_a a canonically conjugate pair of anticommuting ghosts η^a, $\mathcal{P}_a$, obeying

$$[\mathcal{P}_a, \eta^b] = [\eta^b, \mathcal{P}_a] = -\delta_a^b \tag{16}$$

$$(\eta^a)^* = \eta^a, \qquad (\mathcal{P}_a)^* = -\mathcal{P}_a \tag{17}$$

and

$$[\mathcal{P}_a, q^i] = [\mathcal{P}_a, p_i] = [\eta^a, q^i] = [\eta^a, p_i] = 0 \tag{18}$$

The super phase space whose coordinates are $(q^i, p_i, \eta^a, \mathcal{P}_a)$, equipped with this Poisson bracket structure, is called the "extended phase space."†

It is convenient to define an additional structure on the extended phase space, that of "ghost number." This is done by attributing the following ghost number to the canonical variables: the q^i, p_i have ghost number zero, the ghosts η^a have ghost number one, whereas the "antighosts" $\mathcal{P}_a$ have ghost number minus one. Moreover, one requires that the ghost number of a product of variables (with definite ghost number) is equal to the sum of their ghost numbers.

Let us now consider the "generating function"

$$\Omega = \sum_{n \geq 0} \eta^{a_{n+1}} \cdots \eta^{a_1} \overset{(n)}{U}{}^{b_1 \ldots b_n}_{a_1 \ldots a_{n+1}} \mathcal{P}_{b_n} \cdots \mathcal{P}_{b_1} \tag{19}$$

whose derivatives of order $n + 1$ with respect to the ghost, and of order n with respect to the antighosts, are just the nth-order structure functions, e.g.,

$$\left. \frac{\partial \Omega}{\partial \eta^a} \right|_{\eta = \mathcal{P} = 0} = \overset{(0)}{U}_a \tag{20a}$$

$$\left. \frac{\partial^3 \Omega}{\partial \eta^b \partial \eta^a \partial \mathcal{P}_c} \right|_{\eta = \mathcal{P} = 0} = 2 \overset{(1)}{U}{}^c_{ab} \tag{20b}$$

etc. Because of (20a), Ω generates, to first order in the ghosts, a gauge transformation in which the (commuting) "infinitesimal parameter ε^a" is replaced by the (anticommuting) ghost η^a. For that reason, Ω is called the BRST generator [1, 2].

The BRST generator possesses the following properties:

1. Ω is real: $\Omega = \Omega^*$ (21a)

2. Ω has ghost number +1: $\mathrm{gh}(\Omega) = 1$ (21b)

3. Ω is "anticommuting": $\varepsilon(\Omega) = 1$ (21c)

(see Refs. 17 and 11)

† Note that the Poisson bracket is symmetric for anticommuting variables.

More importantly, it is nilpotent:

$$[\Omega, \Omega] = 0 \tag{22}$$

Indeed, one checks that (22) simply follows from the definition (13) of the structure functions. Since we know that the identities (12) that guarantee the existence of the structure functions do hold, we conclude that Ω exists and is nilpotent. [Without these identities, $\overset{(n)}{U}$ would not exist and a nilpotent Ω could not be constructed.]

Hence, one sees that the structure of constrained Hamiltonian systems is completely captured by the BRST generator. The nilpotency property (22), together with (19), is completely equivalent to (12) and (13).

3.2. Rank of a System of Structure Functions

The transformation generated by Ω is called the BRST symmetry. If $\overset{(n)}{U} = 0$, for $n > N$, one says that Ω and the associated system of structure functions is (at most) of rank N. The higher the rank, the more the BRST symmetry "mixes" the ghosts with the original canonical variables.

Because the structure functions are not completely determined by their defining equations, there is also some ambiguity in Ω and in the concept of rank: the rank is associated with a definite choice for Ω (and not with a given constraint surface).

For Abelian theories, the BRST generator can be taken to reduce to

$$\Omega = \eta^a G_a \qquad ([G_a, G_b] = 0) \tag{23}$$

and the rank is zero. For gauge theories of the Yang-Mills type, with a true algebra, Ω becomes

$$\Omega = \eta^a G_a - \tfrac{1}{2}\eta^b \eta^c C^a_{cb} \mathcal{P}_a \tag{24}$$

and is of rank one. Open algebras generically lead to higher ranks (e.g., supergravity [8], the relativistic membrane [18]).

3.3. Off-Shell Nilpotency of Ω

A remarkable feature of the Hamiltonian formulation of the BRST symmetry is that it yields, for any gauge system, a BRST charge that is nilpotent independently of the field equations [(3.7) is an identity throughout phase space]. This is in contrast with the Lagrangian construction of the BRST symmetry, which leads to a BRST charge that is only on-shell nilpotent in the open algebra case. In that sense, the conjugate momenta appear as auxiliary variables, making (3.7) hold off-shell.

The off-shell nilpotency of the BRST charge is extremely important when it comes to discuss the associated cohomology (which can only be defined because $[\Omega, \Omega] = 0$).

4. CANONICAL COVARIANCE OF THE STRUCTURE FUNCTIONS

4.1. Classical Ghosts as Revealing the Canonical Covariance of the Structure Functions

Different choices of structure functions yield different BRST generators. How are these related?

It was already pointed out that two different sets of structure functions are not obtained from one another by a canonical transformation in the original, nonextended phase space.

In view of this negative statement, the following result appears quite surprising [19, 11].

Theorem. Any two BRST generators are related by a canonical transformation in the *extended* phase space.

Accordingly, the ambiguity in the structure functions can be completely accounted for by a mere canonical transformation in the extended phase space. It simply corresponds to different, canonically related realizations of the BRST symmetry.

We will not prove the above theorem here in its full generality (see Refs. 11 and 19). Rather, we will illustrate it in a particular case.

4.2. Any Gauge Theory Can Be (Locally) "Abelianized"

It is a well-known theorem on constrained Hamiltonian systems that every set of first class constraints can be locally replaced by equivalent constraints identically in involution [20, 11],

$$F_a = a_a^b \overset{(0)}{U}_b, \qquad \det a \neq 0 \tag{25}$$

$$[F_a, F_b] = 0 \tag{26}$$

One says that every gauge theory can be "Abelianized" [21] [to get (25), solve for the momenta!].

For the Abelian constraints $F_a = 0$, the construction of the BRST generators is immediate:

$$\Omega^F = \eta^a F_a \tag{27}$$

One has indeed from (26)

$$[\Omega^F, \Omega^F] = 0 \tag{27}$$

Let us perform the extended phase space canonical transformation generated by

$$C = \eta^b \varepsilon_b^a \mathscr{P}_a \tag{28}$$

where $\varepsilon^a_b(q, p)$ are infinitesimal, arbitrary functions. In that transformation, the BRST generator transforms as

$$\delta\Omega = \Omega - \Omega^F = [\Omega^F, C] \tag{29}$$

(to first order). One finds explicitly

$$\delta\Omega = \eta^a \varepsilon^b_a F_b + \mathscr{P}\eta\eta \text{ terms} \tag{30}$$

So, the new η^a term in Ω is now given by

$$\Omega = \eta^a F'_a + \text{“more”} \tag{31a}$$

with

$$F'_a = (\delta^b_a + \varepsilon^b_a) F_b \tag{31b}$$

where $F'_a = 0$ are constraints equivalent to $F_a = 0$, but in general no longer identically in involution. The transformation generated by (28) modifies accordingly the coefficient of η^a in Ω at will within the allowed class.

By repeated applications of canonical transformations of the form (28), one can generate in Ω the desired first term $\eta^a \overset{(0)}{U}_a$,

$$\Omega = \eta^a \overset{(0)}{U}_a + \text{“more”} \tag{32}$$

Since Ω is obtained from Ω^F by a canonical transformation, its nilpotency is automatic.

This analysis indicates:

1. That, indeed, the transition from one set of constraints to an equivalent one can be realized by a canonical transformation in the extended phase space.
2. That a nilpotent Ω of the form (32) exists (and hence, the structure functions introduced above can indeed be defined, as was stated in Refs. 11 and 12, last reference).

5. BRST OBSERVABLES

5.1. Gauge Invariant Functions

Classical "observables" are gauge invariant functions $A_0(q, p)$, i.e., functions that weakly commute with the constraints,

$$[A_0, G_a] \approx 0 \tag{33}$$

Two observables A_0, A_0' are considered as equivalent if they differ by a term that vanishes on the constraint surface

$$A_0' = A_0 + k^a G_a \approx A_0 \tag{34}$$

In that case, they possess the same classical values.

It is easy to show, by using the Jacobi identity, that the set of observables is closed under the Poisson bracket operation (and also, of course, under addition, multiplication, ...). Furthermore, it follows from the first class property of the constraints that if A_0' and B_0' are, respectively, equivalent to A_0 and B_0, then, their Poisson bracket $[A_0', B_0']$ is equivalent to $[A_0, B_0]$. Accordingly, the space of equivalence classes of observables, defined by (34), is naturally equipped with a Poisson bracket structure [22]. Note that in the open case, one cannot find, in general, a representative in each equivalence class that obeys $[A_0, G_a] = 0$ (strongly).

5.2. BRST Invariant Extensions of Classical Observables

An extension A of the observable $A_0(q, p)$ is a function in the extended phase space with the following properties:

1. $A|_{\eta=\mathscr{P}=0} = A_0(q, p)$ (35a)
2. $gh(A) = 0$ (35b)

From (35a) and (35b), it follows that one can expand A as

$$A = \sum_{n\geq 0} \eta^{b_n} \ldots \eta^{b_1} \overset{(n)}{A}{}^{a_1\ldots a_n}_{b_1\ldots b_n} \mathscr{P}_{a_n} \ldots \mathscr{P}_{a_1} \tag{36}$$

with $\overset{(0)}{A} \equiv A_0$.

It is clear that if $[A, \Omega] = 0$, then, $[\overset{(0)}{A}, G_a] \approx 0$: BRST invariant functions are extensions of gauge invariant observables.

The converse is also true: any gauge invariant function $A_0(q, p)$ possesses an appropriate extension that is BRST invariant,

$$[A, \Omega] = 0 \tag{37}$$

Furthermore, the extension is not unique: two extensions A and A' differ by

$$A' = A + [K, \Omega] \tag{38}$$

where K has ghost number -1.

The proof of this theorem will not be reproduced here. It can be found in Refs. 11 and 12. Let us simply make the result plausible by noticing the following: (i) By virtue of the Jacobi identity, and the nilpotency of Ω, $[K, \Omega]$ is BRST invariant:

$$[[K, \Omega], \Omega] = 0 \tag{39}$$

(ii) After the theory has been "Abelianized," one can adjust k^a in (34) so that $[A_0, F_a] = 0$ (strongly) [20, 11]; then the theorem is obvious ($A = A_0$ is a permissible extension). (iii) With $K = k^a \mathscr{P}_a$, $[K, \Omega^F]$ takes the form $k^a F_a +$ "more", which is just the ambiguity in A_0 mentioned above.

5.3. BRST Observables

Any function A in the extended phase space possessing the properties (35) and (38) is called a BRST observable. Two BRST observables differing by $[K, \Omega]$, for some K, must be identified.

One easily checks that the space of equivalence classes of BRST observables is naturally equipped with a Poisson bracket structure. Furthermore, we just indicated that there is a bijective correspondence between equivalence classes of gauge-invariant observables and BRST observables.

5.4. BRST Invariant Extensions of the Constraints

Because of the first class property, the zeroth-order structure functions are gauge invariant. One can hence associate with them BRST invariant extensions. One possible choice is

$$G_a \to \bar{G}_a = [-\mathscr{P}_a, \Omega] \tag{40}$$

This choice explicitly shows that the BRST invariant extensions of the constraints belong to the same equivalence class as the observable 0 (i.e., $\bar{G}_a \sim 0$).

One interesting property of the extensions $\bar{G}_a$ is that they obey the same algebra as the original constraints

$$[\bar{G}_a, \bar{G}_b] = C^c_{\ ab} \bar{G}_c \tag{41}$$

provided the first-order structure functions are constant (closed algebra).

However, in the open algebra case, there is in general no way to fulfill (5.9), for $C^c_{\ ab} \bar{G}_c$ is not BRST invariant,

$$\begin{aligned}[C^c_{\ ab} \bar{G}_c, \Omega] &= [C^c_{\ ab}, \Omega] \bar{G}_c \\ &= [C^c_{\ ab}, G_d] \bar{G}_c \eta^d + \text{higher-order terms} \\ &\neq 0\end{aligned}$$

(and $C^c_{\ ab}$ does not even possess a BRST invariant extension for it is not gauge invariant in general).

Hence, it is felt that one should not attach too much importance to the particular result (41). The key object to look at is not the set of BRST invariant extensions $\bar{G}_a$ of the constraints, but rather, the BRST generator itself.

5.5. Trajectories in Extended Phase Space

Like all other observables, the Hamiltonian H_0 possesses a BRST invariant extension H,

$$[H, \Omega] = 0 \tag{42}$$

This extension is determined up to

$$H \to H + [K, \Omega] \tag{43}$$

As shown by Fradkin and his collaborators, a change of gauge in the path integral amounts to a transformation like (43). Hence, in order to enforce gauge invariance in the quantum domain, we must define the physical subspace so that the change (43) is physically irrelevant.

The trajectories in the extended phase space are generated by H,

$$\frac{dA}{dt} = [A, H] \tag{44}$$

If the ghost are initially zero,

$$\eta^a(t = 0) = \mathcal{P}_a(t = 0) = 0 \tag{45}$$

they remain so because H has zero ghost number. The trajectories (44) project then "down" to the trajectories in the unextended phase space generated by H_0.

6. QUANTUM THEORY—FORMAL CONSIDERATIONS

6.1. Quantum Linear Space of States

In the quantum theory, the canonical variables and the ghosts become operators in a linear space with a nonpositive inner product. This inner product is such that the real canonical variables become Hermitian operators, whereas the imaginary ones become anti-Hermitian. As a result, both the quantum BRST generator Ω and the quantum BRST Hamiltonian H are Hermitian.

We will assume throughout that one can order the BRST generator Ω in such a way that it is not only Hermitian, but also nilpotent:

$$[\Omega, \Omega] = 2\Omega^2 = 0 \tag{46}$$

Note that these two requirements ($\Omega^2 = 0, \Omega^* = \Omega$) are compatible only with an indefinite metric. Also, we will assume

$$[H, \Omega] = 0 \tag{47}$$

Quantum observables are zero ghost number operators A which commute with Ω, i.e., which are BRST invariant,

$$[A, \Omega] = 0 \tag{48}$$

This definition is just the quantum transcription of the corresponding classical concept.

6.2. Ghost Number Operator

The ghost number of an operator is defined as in the classical theory. It can be represented by the operator

$$Q_c = i\eta^a \mathcal{P}_a + \text{const} \tag{49}$$

which obeys

$$[Q_c, p_i] = [Q_c, q^i] = 0 \tag{50a}$$

$$[Q_c, \eta^a] = \eta^a \qquad [Q_c, \mathcal{P}_a] = -\mathcal{P}_a \tag{50b}$$

If desired, one can adjust the constant in Q_c so that Q_c is anti-Hermitian†:

$$Q_c = \frac{i}{2}(\eta^a \mathcal{P}_a - \mathcal{P}_a \eta^a), \qquad Q_c^* = -Q_c \tag{51}$$

It follows from (6.5) that if a state $|f\rangle$ has definite ghost number $k(Q_c|f\rangle = k|f\rangle)$, then, $q^i|f\rangle$ and $p_i|f\rangle$ have also ghost number k, whereas $\eta^a|f\rangle$ and $\mathcal{P}_a|f\rangle$ have ghost numbers $k+1$ and $k-1$, respectively.

Although anti-Hermitian, Q_c possesses real eigenvalues. This can be seen by expanding the states in the η^a representation as

$$|\psi\rangle = |\psi\rangle^D + |\psi_a\rangle^D \eta^a + |\psi_{ab}\rangle^D \eta^a \eta^b + \cdots \tag{52}$$

Here, $|\psi\rangle^D, |\psi_a\rangle^D \ldots$ are states that live in the Hilbert space of the original variables q^i, p_i (they do not involve the ghosts). [See Refs. 23 and 11 for more on that representation. Note that in Ref. 11, the constant in (49) was naturally adjusted so that the first term in (52) has ghost number zero.] With (51), $|\psi\rangle^D$ has ghost number $-m/2$, $|\psi_a\rangle^D \eta^a$ has ghost number $-m/2 + 1 \ldots$ and the last term in the expansion (52) has ghost number $m/2$.

There is no contradiction between anti-Hermiticity of Q_c and real eigenvalues, because the eigenvectors of Q_c have zero norm‡ (except, maybe, those associated with the eigenvalue zero, which appears when m is even). Hence, the presence of zero norm eigenstates when the ghost number is half-integer ("fractionalization of the ghost number" m odd) is inescapable and reflects a major feature of the BRST formalism.

† Q_c differs here from the Q_c of Ref. 9 by the factor i.

‡ If that norm is not infinite, in which case one cannot talk of eigenvector!

6.3. Physical Subspace

The pseudo-Hilbert space considered so far contains spurious degrees of freedom: those associated with the ghosts, and those corresponding to the pure gauge variables among the (q^i, p_i). In order to get a satisfactory theory, one must discard these spurious degrees of freedom by means of a subsidiary condition on the accessible physical states.

The subsidiary condition must obey the following requirements:

1. It must be linear, i.e., it must select a subspace.
2. BRST observables should map the physical subspace on itself.
3. Trivial observables of the form $[K, \Omega]$ should have vanishing matrix elements between physical states, since, as we have seen, they must be identified with zero if the quantum theory is to be gauge invariant.

This suggests adopting [9, 10]

$$\Omega|\psi\rangle = 0 \tag{53}$$

as definition of the physical subspace: physical states are annihilated by the BRST charge.

That this definition is appropriate will be seen in Section 7, where it will be shown that with (53), the BRST quantum theory is equivalent to other approaches to quantization, in simple cases, at least.

It is easy to verify that (53) fulfills all three requirements 1-3. The verification of 3 uses in an essential way the Hermiticity of Ω. I would thus be extremely reluctant in working with a scalar product such that $\Omega^* \neq \Omega$.

6.4. Quantum Gauge Invariance

With (53), one can identify quantum BRST observables differing by $[K, \Omega]$,

$$A \sim A + [K, \Omega] \tag{54}$$

Indeed, when acting on physical states, $[K, \Omega]$ creates null states that decouple from all other physical states.

To the identification (54) of observables corresponds the identification of physical states according to

$$|\psi\rangle \sim |\psi\rangle + \Omega|\chi\rangle \tag{55}$$

where $|\chi\rangle$ is an arbitrary state. This is because (i) $\Omega|\chi\rangle$ is a physical state since $\Omega^2 = 0$; (ii) $\Omega|\chi\rangle$ decouples from all other physical states. The true physical subspace is thus the quotient space of (53) by (55) (="$\text{Ker}\,\Omega/\text{Im}\,\Omega$"). The BRST cohomology is therefore fundamental.

The transformations (54) and (55) are called "quantum gauge transformations." This is rightly so because (54) guarantees that one can replace H by $H+[K,\Omega]$ in the evolution operator $\exp -iHt$, without changing transition amplitudes (this replacement corresponds to a change of gauge in the path integral as indicated by Fradkin and his school). Furthermore, the quotient relation (55) kills enough states from (53) so as to make the number of degrees of freedom in the quantum and classical theories equal to each other (in the simplest cases). Without this property, gauge invariance would not be realized quantum mechanically.

It should be clear that the quantum gauge invariance (54), (55) would not exist if Ω was not nilpotent (and Hermitian). Hence, one can say that the nilpotency of Ω is the quantum expression of the gauge invariance principle. Without $\Omega^2=0$, one could not go from one gauge to another, and gauge invariance would be lost at the quantum level. [It is not completely obvious, however, that this would imply inconsistency of the theory.]

It should be noted here that there may be an anomaly in the algebra of the quantum constraints $G_a(q,p)$ without anomaly in the BRST algebra $\Omega^2=0$ (and in the algebra of the BRST invariant extensions $\bar{G}_a$). This result holds, for instance, in string theory, where the ghosts also contribute anomalous terms in Ω^2. These cancel the anomalous contributions from the original degrees of freedom exactly in the "physical" dimension [3]. It thus appears that the BRST approach is more flexible than other, more traditional methods of quantization that do not involve the ghosts.

Finally, we note that the "gauge transformations" generated by the BRST invariant extensions $\bar{G}_a$ of the constraints are just a subset of (55). Indeed, $\bar{G}_a=[-\mathcal{P}_a,\Omega]$ creates null states. The converse, however, is not true: (55) is a much bigger gauge invariance because $|\chi\rangle$ in (55) can be an arbitrary $|\psi\rangle$-dependent state. The "quantum gauge invariance" (55) seems thus to be more fundamental.

6.5. Unitarity in the Physical Subspace

Because the Hamiltonian H is an observable, the evolution operator $\exp(-iHt)$ commutes with the BRST charge Ω. Moreover, it is unitarity since H is Hermitian. This means that the time evolution leaves the physical space invariant. This also implies unitarity in the physical subspace. If the scalar product $\langle\psi_1|\psi_2\rangle$ of two physical states is positive definite once the zero norm states have been factored out, one manifestly gets a satisfactory quantum theory. This happens in the usual cases. Otherwise, some additional measure has to be taken (see Section 8).

We have assumed here that there were no "anomalous" terms in $[H,\Omega]=0$. In the presence of such terms, the evolution operator would map a physical state out of the physical subspace. This leads to difficulties.

Note finally that the transition amplitudes $\langle\psi_1|\exp(-iHt)|\psi_2\rangle$ are invariant under the replacement of H by $H + [K, \Omega]$ even if K does not possess ghost number minus one. This simply follows from the fact that $|\psi_1\rangle$ and $|\psi_2\rangle$ are annihilated by Ω. Hence, $\langle\psi_1|[K, \Omega]|\psi_2\rangle = 0$ for arbitrary K. When $gh(K) \neq -1$, the ghost number is not conserved, $[Q_c, H] \neq 0$, but this does not affect the physical sector.

6.6. Anomalies

The previous discussion has indicated the importance of the quantum mechanical conditions

$$\Omega^2 = 0 \tag{56}$$

$$[H, \Omega] = 0 \tag{57}$$

It might be that (56) or (57) is not actually realized owing to quantum effects. This occurs, e.g., in string theory away from the critical dimension [(56) does not hold]. As to the Yang-Mills anomalies, they seem to be related to a violation of (57) and probably also to (56), although the problem is far from being completely cleared.

When neither (56) or (57) is true at the quantum level, the quantum theory seems to be inconsistent.

7. EQUIVALENCE OF BRST METHODS WITH OTHER APPROACHES TO QUANTIZATION IN SIMPLE CASES

7.1. Abelian Constraints, Schrödinger Representation

In view of the local "Abelianalizability" of the constraints, a natural case to look at in order to understand the condition (53) is when the constraints are pure momenta,

$$G_a = p_a, \qquad [G_a, G_b] = 0 \tag{58}$$

Then the coordinates (q^i, p_i) split into two groups (q^α, p_α) and (q^a, p_a) $(\alpha = 1, \ldots, n-m)$.. The variables q^α, p_α are true, gauge invariant degrees of freedom, whereas (q^a, p_a) are pure gauge. The BRST generator reads

$$\Omega = \eta^a p_a \tag{59}$$

and a general BRST state is given by

$$|\psi\rangle = |\psi\rangle^D + |\psi_a\rangle^D \eta^a + |\psi_{ab}\rangle^D \eta^b \eta^a + \cdots \tag{60}$$

with $|\psi\rangle^D = \psi(q^\alpha, q^a)$ in the Schrödinger representation.

Let us compute $\Omega|\psi\rangle$. One finds

$$\Omega|\psi\rangle = \frac{\hbar}{i}\left[\partial_a \psi^D \eta^a + \tfrac{1}{2}(\partial_b \psi_a - \partial_a \psi_b)\eta^b \eta^a + \cdots\right] \tag{61}$$

This shows that the BRST operator is the exterior derivative operator d in the space of the q^a. BRST states can be viewed as forms. The ghost number of a state, adjusted as in Ref. 11, is just the rank of the form. From (61), one sees that a physical state, rewritten as

$$|\psi\rangle = \overset{(0)}{\psi} + \overset{(1)}{\psi} + \overset{(2)}{\psi} + \cdots \tag{62}$$

($\overset{(0)}{\psi}$ = first term in (60) = 0-form, etc...), must be a closed form

$$d\overset{(0)}{\psi} = d\overset{(1)}{\psi} = d\overset{(2)}{\psi} = \cdots = 0 \tag{63}$$

Furthermore, adding to $|\psi\rangle$ the state $\Omega|\chi\rangle$ amounts to modifying $\overset{(0)}{\psi}, \overset{(1)}{\psi}, \overset{(2)}{\psi}, \ldots$ as

$$\begin{aligned} \overset{(0)}{\psi}{}' &= \overset{(0)}{\psi} \\ \overset{(1)}{\psi}{}' &= \overset{(1)}{\psi} + d\overset{(0)}{\chi} \\ \overset{(2)}{\psi}{}' &= \overset{(2)}{\chi} + d\overset{(1)}{\chi} \\ &\text{etc.} \end{aligned} \tag{64}$$

Accordingly, the physical subspace is just given by Ker $d/$Im d (in the q^a space).

If the topology of the q^a space is trivial, one can set $\overset{(1)}{\psi}, \overset{(2)}{\psi}$ and all higher-order terms equal to zero by an appropriate choice of $|\chi\rangle$. This means that one can take a representative in each equivalence class of physical states with zero ghost number (as adjusted in Ref. 11). So, the requirement of zero-ghost number is not a further assumption, but rather is a "gauge condition" on the "quantum gauge invariance."

For zero ghost number representatives, $|\psi\rangle = \overset{(0)}{\psi}(q^\alpha, q^\alpha)$, the BRST condition reads

$$\frac{\hbar}{i}\frac{\partial \overset{(0)}{\psi}}{\partial q^a} = 0 \tag{65}$$

$\overset{(0)}{\psi}$ must thus be independent of q^a. These are exactly the physical state conditions of the Dirac approach.

This result establishes the equivalence between the BRST and Dirac methods of quantization. Although $\Omega|\psi\rangle$ is a single condition, the quantum gauge invariance, without further requirement, implies a one-to-one correspondence between the physical states in both approaches. (One can also establish that the observables and the scalar products coincide, but this requires some regularization [11].)

If the topology of the q^a space is nontrivial, these results may no longer hold: a doubling of BRST states can occur [15]. This problem is examined in Section 8.

7.2. Abelian Constraints—Fock Space Representation

In electromagnetism, the BRST generator reads

$$\Omega = \int d^3x\,(\nabla \cdot \mathbf{E}\, C - i\mathcal{P}\pi^0) \tag{66}$$

where $\mathbf{E}$ is the electric field, conjugate to the spatial components of the potential, π^0 is the momentum conjugate to A_0, whereas C is the ghost and $\mathcal{P}$ the momentum conjugate to the antighost [11]:

$$[C(x), \bar{\mathcal{P}}(x')] = -\delta(x, x') \tag{67a}$$

$$[\bar{C}(x), \mathcal{P}(x')] = -\delta(x, x') \tag{67b}$$

$$[A_0(x), \pi^0(x')] = \delta(x, x') \tag{67c}$$

$$[A_i(x), E_j(x')] = \delta_{ij}\delta(x, x') \tag{67d}$$

All other brackets vanish.

Expansion of the fields in Fourier modes leads, after some rearrangements, to

$$\Omega = i\int d^3k[C^*(k)a(k) - a^*(k)C(k)] \tag{68}$$

where one has the following commutation/anticommutation relations:

$$[a(k), a^*(k')] = 0 = [C(k), C^*(k')] \tag{69}$$

$$[C(k), \bar{C}^*(k')] = \delta(k, k') = [\bar{C}(k), C^*(k')] \tag{70}$$

The operators $a(k)$, $a^*(k)$ are related to $a^\mu(k)$, $a^{*\mu}(k)$ as (with an appropriate definition of a^μ, $a^{*\mu}$ in terms of A_μ, π^μ)

$$a(k) = a^0(k) - a^{\text{long}}(k)$$

$$a^*(k) = a^0(k)^* - [a^{\text{long}}(k)]^*$$

where $a^{\text{long}}(k)$ is the longitudinal part of $a^i(k)$. They fulfill (69) as a result of

$$[a^\mu(k), a^{*\nu}(k')] = \eta^{\mu\nu}\delta(k, k') \tag{71}$$

We want to show that any physical state solution to $\Omega|\psi\rangle = 0$ can be written as

$$|\psi\rangle = |P\rangle|0\rangle_{\text{temporal}}|0\rangle_{\text{longitudinal}}|0\rangle_{\text{ghost}} + \Omega|\chi\rangle \tag{72}$$

i.e., that there is in each equivalence class of physical states a representative that is annihilated by the ghost destruction operators

$$C(k)|\psi\rangle = \bar{C}(k)|\psi\rangle = 0 \tag{73}$$

as well as by the temporal and longitudinal photon destruction operator

$$a^0(k)|\psi\rangle = a^{\text{long}}(k)|\psi\rangle = 0$$
$$\Updownarrow$$
$$a(k)|\psi\rangle = b(k)|\psi\rangle = 0 \tag{74}$$

Here $b(k)$ is given by

$$b(k) = a^0(k) + a^{\text{long}}(k) \tag{75a}$$

and is canonically conjugate to $a(k)$

$$[b(k), a^*(k)] = -2 \tag{75b}$$

In (72), $|P\rangle$ is a purely transverse state.

The BRST approach appears thus to be very close to the Gupta-Bleuler method of quantization of the electromagnetic field.

It should be stressed again that, as a result of (72), the conditions (73) and (74) are not additional requirements on the physical states, but rather are gauge fixing conditions for the quantum gauge invariance (55). Also, note that the ghost number operator reads, in terms of Fourier components,

$$Q_c = -\int dk(C_k^*\bar{C}_k - \bar{C}_k^*C_k) \tag{76}$$

The states obeying (73) have zero ghost number, and hence there is no contradiction with the fact that they possess a positive norm. The decomposition (72) is known as the "Kugo-Ojima" quartet mechanism for the elimination of ghosts [9]. To prove it, let us discretize the modes and write them as a_n with n an integer $[a(k) \to a_n]$.

Consider the new operator, R,

$$R = \sum_n [C_n^*\bar{C}_n + \bar{C}_n^*C_n - \tfrac{1}{2}(a_n^*b_n + b_n^*a_n)] \tag{77}$$

This operator is Hermitian and counts the modes of the ghosts, of the temporal photons, and of the longitudinal photons. It can be diagonalized.

Its eigenstates are given by the monomials

$$|\{\lambda_n, \mu_n, \alpha_n, \beta_n\}, f\rangle = \prod_k \bar{c}_k^{*\lambda_k} c_k^{*\mu_k} a_k^{*\alpha_k} b_k^{*\beta_k} |f\rangle \tag{78}$$

where $|f\rangle$ is any state not involving $\bar{c}_n^*$, c_n^*, a_n^*, or b_n^*. One has

$$R|\{\lambda_n, \mu_n, \alpha_n, \beta_n\}, f\rangle = \sum_k (\lambda_k + \mu_k + \alpha_k + \beta_k)|\{\lambda_n, \mu_n, \alpha_n, \beta_n\}, f\rangle \tag{79}$$

The operator R commutes with Ω and can thus be assumed to be diagonal for physical states. Let us therefore suppose

$$R|\psi\rangle = r|\psi\rangle \tag{80}$$

with r a positive integer.

One easily checks that R is a null operator,

$$R = [K, \Omega] \tag{81}$$

with

$$K = \frac{i}{2}(\sum_n b_n^* c_n - c_n^* b_n) \tag{82}$$

Hence, if r in (80) is different from zero, $|\psi\rangle$ is a null state

$$|\psi\rangle = \frac{1}{r} R|\psi\rangle = \frac{1}{r}(K\overset{0}{\cancel{\Omega}} + \Omega K)|\psi\rangle$$

$$|\psi\rangle = \Omega\left(\frac{K}{r}|\psi\rangle\right) \tag{83}$$

Thus, the dependence on the ghosts, on the temporal photons, and on the longitudinal photons of a general physical state can only occur through null states as in (72).

7.3. Inner Products

The Fock representation of Section 7.2 and the Schrödinger representation of Section 7.1 are not equivalent because of scalar product questions.

In Section 7.1 it was implicitly assumed that the inner product was positive definite in the "bosonic sector" involving all the q^i's $[(\psi, g) = \int dq^i\, d\eta\, \psi^* g$; see Ref. 11]. Also, as pointed out in Ref. 15, one was rather sloppy when applying the BRST gauge transformation $|\psi\rangle \to |\psi\rangle + \Omega|\chi\rangle$ (unprecised falloff of $|\psi\rangle$ and $|\chi\rangle$ at infinity). As a result, the norm of the physical states had to be regularized [11]. In Section 7.2, the bosonic oscillator a^0 creates negative norm states. But physical states have a *finite*, positive norm.

The virtue of the Fock representation is its greater rigor. However, it is not applicable to an odd number of constraints; it can only be applied to an even subset. Indeed, *two* pure gauge bosonic variables (a^0 and a^{long}) and *two* ghosts are needed for the mutual cancellation mechanism of the Kugo-Ojima quartet. (Of course one can always make the number of constraints even by adding appropriate pure gauge degrees of freedom. But the grouping of the constraints in pairs with oscillatorlike variables, as required by the quartet mechanism, might not be possible.)

8. DOUBLING OF BRST STATES

When the topology of the configuration space q^i is nontrivial, or for more complicated constraints as in the string model (see Appendix), the comparison between the BRST physical states and the physical states of more conventional approaches is less immediate.

We will not go through the details here [15, 3, 24] but only summarize the main features.

One finds that a general solution to the BRST physical condition can generically be written as

$$|\psi\rangle = |P_1\rangle|0\rangle_{\text{ghost}} + |P_2\rangle|0\rangle_{\text{ghost}}\eta^0 + \Omega|\chi\rangle \tag{84}$$

where (i) $|P_1\rangle$ and $|P_2\rangle$ are traditional physical states not involving the ghosts; (ii) $|0\rangle_{\text{ghost}}$ is the ghost vacuum for the ghosts that decouple by the mechanisms described in Section 7; (iii) η^0 and its conjugate momentum $\mathcal{P}_0$ are ghosts that are not eliminated by the BRST gauge invariance.

Typically, $(\eta^0, \mathcal{P}_0)$ are "zero modes" obeying

$$[\eta^0, \mathcal{P}_0] = 1, \qquad (\eta^0)^* = \eta^0, \qquad (\mathcal{P}_0)^* = \mathcal{P}_0 \tag{85}$$

The irreducible representation space for (85) is well known and given by functions of one Grassmann variable η^0,

$$f = a + b\eta^0 \tag{86a}$$

$$(f, g) = \int f^* g \, d\eta^0 \tag{86b}$$

$$\mathcal{P}_0 = \frac{\partial}{\partial \eta^0} \tag{86c}$$

The states $f_1 = 1$ and $f_2 = \eta^0$ diagonalize the ghost number operator and have zero norm (their mutual scalar product is one). Alternatively, one can consider the states $2^{-1/2}(1 + \eta^0)$, which are orthogonal to each other and of norm ± 1. These latter states do not diagonalize the ghost number operator.

The equation (84) indicates that there are twice as many states in the BRST approach as there are states in the traditional methods ("doubling of states"). The "zero modes" do not "decouple" like the other modes. With more than one "zero mode pair," there would be more than one "doubling."

8.1. Selection Rule

Although a doubling of states appears in the BRST approach, this is of no physical relevance if the operators under consideration are multiples of the identity in the η^0 space,

$$A \sim A' \otimes I \Leftrightarrow A = A' \otimes I + [K, \Omega] \tag{87}$$

Here, A' acts on everything but η^0, and I is the unit operator in the two-dimensional space (86a).

There exist BRST invariant observables that are not of the form (87). But, as argued in Ref. 15, such operators are not encountered in practice—at least within the context of the "first quantized" theory. With the condition (87), the zero modes effectively decouple. Without it, they would not.

When the "selection rule" (87) holds for all "interesting operators," one can restrict one's attention to one sector of BRST physical states isomorphic to the physical space of, say, the Dirac approach. This can be achieved by imposing one further condition ("truncation") on the physical states. The choice of that condition (and the need for it) seems to be to some extent a matter of taste, since the dynamics is identical in each sector.

For instance, since negative norm states should be discarded, one natural choice is to impose $|p_1\rangle = |p_2\rangle$ in (84) [15]. With this condition, (84) becomes

$$|\psi\rangle = |p\rangle|0\rangle_{\text{ghost}}(1 + \eta^0) + \Omega|\chi\rangle \tag{88}$$

The states (88) all have positive norm (given by $\langle p|p\rangle$).

8.2. Diagonal Ghost Number Versus Hermiticity of BRST Charge

The states (88) do not diagonalize the ghost number charge. This has led some people to consider instead states with no component along η^0

$$|\psi\rangle = |p\rangle|0\rangle_{\text{ghost}} \tag{89}$$

These states possess zero norm, something to be expected in view of the properties of the ghost number operator.

To remedy this state of affairs, one can modify the scalar product. This, however, has the following consequences: (i) Ω is no longer Hermitian and arbitrary states of the form $\Omega|\chi\rangle$ no longer decouple; (ii) the ghost number operator is no longer anti-Hermitian.

By adding to Q_c an appropriate real constant one can make it anti-Hermitian in the new scalar product. But then the physical states (89) under consideration, which have positive norm, must possess zero new ghost number. "Fractionalization" of the ghost number, if any, is lost in the new scalar product.

It is not clear whether these features are desirable and whether (89) should be preferred over (88).

It appears that these different points of view can be made to actually agree if one adopts the following interpretation.

To a large extent, all the sectors are equivalent, since the evolution operator and the S matrix (viewed as an operator, not as a collection of matrix elements) are identical in each sector (provided the factorization condition $H = H' \otimes I$ holds). Hence, for the purpose of determining the S-matrix operator, say, one can work in any sector, even in one that contains only zero norm states if this turns out to be more convenient.

Care need only be taken when one computes transition amplitudes, which involve scalar products. If the S-matrix operator is known in a zero norm sector, one rotates it into a positive norm one (where it is identical!) before computing amplitudes.

Technically, this "rotation" might appear to be the same as a change of scalar product. But it is the author's belief that conceptually it is important to make sure that everything can be phrased without changing the scalar product. Hermiticity of the BRST generator is indeed an essential element of the BRST formalism, for it guarantees the decoupling of the null states.

ACKNOWLEDGMENTS. The author is grateful to Claudio Teitelboim for many useful discussions and for his kind invitation to the meeting on Quantum Mechanics of Fundamental Systems. This work has been supported in part by a travel grant from the Foundation "Agathon De Potter" (Académie Royale de Belgique).

APPENDIX: BRST PHYSICAL STATES IN THE STRING MODEL

We discuss here the "doubling phenomenon" in the case of the string model. For simplicity, we study the bosonic open string, but similar considerations apply to the other models as well (e.g., in the closed bosonic case, one finds two "doublings", etc.). Only the first quantized theory is considered.

Our presentation closely follows the first reference in Ref. 3, to which we refer for the details, with a slight improvement, namely, we only impose the BRST condition on the physical states. The extra condition $(L-1)|\psi\rangle = 0$ need not be imposed by hand. It indeed follows from the BRST formalism that if a physical state is not annihilated by $L-1$, then it is a null state ($|\psi\rangle = \Omega|\chi\rangle$). This is the content of the first theorem below.

The BRST charge reads [3]

$$\Omega = \eta^0(L-1) - \mathcal{P}_0 M + \bar{\Omega} \tag{A.1}$$

where we have explicitly written out the ghost zero mode dependence.

The operators M and $\bar{\Omega}$ do not involve $\mathcal{P}_0$, η^0 and have the following properties [3]: (i) M annihilates the ghost vacuum,

$$M|0\rangle_{\text{ghost}} = 0 \tag{A.2}$$

Here, $|0\rangle_{\text{ghost}}$ is the state annihilated by all ghost destruction operators c_n, $\bar{c}_n$ with *strictly* positive n ($n > 0$). (ii) $\bar{\Omega}$ is nilpotent in the subspace $L = 1$. Moreover, any state in that subspace obeying $\bar{\Omega}|b\rangle = 0$ can be written as [3]

$$|b\rangle = |p\rangle|0\rangle_{\text{ghost}} + \bar{\Omega}|c\rangle \tag{A.3}$$

where $|P\rangle$ is a physical state of the usual covariant approach [$(L_0 - 1)|P\rangle = 0$, $L_n|P\rangle = 0$, $n > 0$; $|P\rangle$ does not contain the ghosts]. Since $\bar{\Omega}$ does not involve the ghost zero mode, we have assumed in (A.3) that $|b\rangle$ is independent of η^0.†

The operator L appearing in (A.1) is given by

$$L = \alpha' p^2 + N^{\text{bosons}} + N^{\text{ghosts}} \tag{A.4}$$

It is the BRST invariant extension of L_0 and as such commutes with Ω. The operator N^{bosons} counts the "boson mass level" ($N^{\text{bosons}} = \sum n a_n^* a_n$), while N^{ghosts} counts the "ghost mass level." It is easy to see that L can be diagonalized in the subspace $\Omega|\psi\rangle = 0$ (no problem arises because of the indefinite metric). Hence, we can work in a given eigenspace of L.

Theorem. The solutions of $\Omega|\psi\rangle = 0$ with $L \neq 1$ are pure gauge, i.e., can be written as $|\psi\rangle = \Omega|\chi\rangle$.

Proof. Suppose $L \neq 1$. Expand $|\psi\rangle$ as $|\psi\rangle = |a\rangle + |b\rangle\eta^0$, where $|a\rangle$ and $|b\rangle$ are states that do not involve the ghost zero modes. Because $L \neq 1$, one can add to $|\psi\rangle$ a state $\Omega|\chi\rangle$ such that $|b\rangle = 0$. Take, e.g., $|\chi\rangle = -(L-1)^{-1}|b\rangle$. With $|b\rangle = 0$, the BRST condition on $|\psi\rangle$ implies

$$\bar{\Omega}|a\rangle = 0, \qquad (L-1)|a\rangle = 0 \tag{A.5}$$

from which one infers $|a\rangle = 0$, as asserted.

† The pseudo-Hilbert space at hand is the direct product of the string center-of-mass space, the bosonic oscillator Fock space, the ghost Fock space ($n > 0$), and the zero mode space. The momentum p^μ commutes with Ω and can be assumed to be diagonal.

This theorem enables one to consider only the case when

$$(L-1)|\psi\rangle = 0 \tag{A.6}$$

to which we now turn.

When (A.6) holds, $\bar{\Omega}$ is nilpotent and the BRST equation reads, with

$$|\psi\rangle = |a\rangle + |b\rangle\eta^0 \tag{A.7}$$

$$\bar{\Omega}|b\rangle = 0, \qquad \bar{\Omega}|a\rangle - M|b\rangle = 0 \tag{A.8}$$

Theorem. The general solution to $\Omega|\psi\rangle = 0$ can be written as

$$|\psi\rangle = |P_1\rangle|0\rangle_{\text{ghost}} + |P_2\rangle|0\rangle_{\text{ghost}}\eta^0 + \Omega|\chi\rangle \tag{A.9}$$

where $|P_1\rangle$ and $|P_2\rangle$ are purely bosonic states of the usual covariant approach $[L_n|P_i\rangle = 0 = (L_0-1)|P_i\rangle,\ n>0]$.

Proof. From $\bar{\Omega}|b\rangle = 0$ and (A.3), one can assume that the $|b\rangle$ term in $|\psi\rangle$ has the required form. But then, $M|b\rangle = 0$ and the second condition (A.8) reduces to $\bar{\Omega}|a\rangle = 0$. A second application of (A.3) easily leads to the theorem.

This theorem is interesting, because it completely characterizes the physical subspace of the covariant approach. Let us stress again that $|P_2\rangle$ obeys the mass shell condition $(L_0-1)|P_2\rangle = 0$ as a result of the first theorem, even though one might naively think that this property is lost when writing out $\Omega|P_2\rangle|0\rangle_{\text{ghost}}\eta^0 = 0$. If $(L_0-1)|P_2\rangle \neq 0$, $|P_2\rangle$ is a null state which can be removed.

Note. In the case of the open string model, the following remark should be added concerning the doubling phenomenon. The physical states have actually an infinite norm if one takes into account the integration over the space-time momentum $\int d^{26}p$ in the scalar product. (This integration is often treated separately.) Indeed, because of the zero mode constraint $(L_0 - 1)|P\rangle = 0$ ("on the mass-shell" condition) one can only consider wave packets with definite mass. Thus, one picks up a factor of $\delta(0)$ in the scalar product of states belonging to the same mass level through the integration over the mass. This infinity must be regularized.

As argued in Ref. 11, one cannot separate the integration over the gauge degrees of freedom, responsible for the occurrence of $\delta(0)$, from the integration over the ghosts. Gauge degrees of freedom and ghosts are "BRST supersymmetric partners." One way to regularize the scalar product with this remark in mind is then to insert an appropriate function $f_\varepsilon\ (p_\mu, \eta^0)$ (with $f_\varepsilon \to 1$ for $\varepsilon \to 0$) in the relevant integrals [11]. With such a factor, one finds that the norm of a physical state is given by

$$\langle\psi|\psi\rangle = \delta(0)[(P_1|P_2) + (P_1|P_2)] + (P_1|P_1) \tag{A.10}$$

where (P_i, P_j) stands here for the Fock space scalar product without integration over the momenta. Loosely speaking, the term $(P_1|P_1)$ survives—and is finite—because it is multiplied in $\langle\psi|\psi\rangle$ by $\delta(0)$ (mass integral) times 0 ($\delta\eta^0$ integral). This ill-defined product of bosonic and fermionic delta functions has been regularized to 1 in Ref. 11.

For $\langle\psi|\psi\rangle$ to make sense, one must impose the condition $|P_2\rangle = 0$ so as to get rid of the $\delta(0)$-infinite term. This procedure eliminates the doubling and in that sense, appears natural. Being based on a particular though natural regularization of ill-defined expressions, it might, however, not be satisfactory to everyone.

In the case of the monopole system treated in Ref. 15, states occurring in the doubling have finite norm. There is accordingly no possibility of eliminating this phenomenon by scalar product arguments. Similarly, one of the closed string doublings cannot be eliminated by similar considerations.

ACKNOWLEDGMENT. This research was supported by the Fonds National Belge de la Recherche Scientifique.

REFERENCES

1. C. Becchi, A. Rouet, and R. Stora, *Phys. Lett. 52B*, 344 (1974).
2. I. V. Tyutin, Gauge invariance in field theory and in statistical mechanics in the operator formalism, Lebedev preprint FIAN N°39 (1975), unpublished.
3. M. Kato and K. Ogawa, *Nucl. Phys. B212*, 443 (1983); S. Hwang, *Phys. Rev. D28*, 2614 (1983).
4. J. H. Schwarz, Fadeev–Popov ghosts and B.R.S. symmetry in string theories, Caltech-preprint CALT-68-1304 (1985); J. Fisch, Quantification BRS des modèles de cordes, Memoire de licence, Université Libre de Bruxelles, Academic Year 1985–1986. See also Ref. 24.
5. E. S. Fradkin and G. A. Vilkovisky, *Phys. Lett. 55B*, 224 (1975); CERN Report Th-2332 (1977).
6. I. A. Batalin and G. A. Vilkovisky, *Phys. Lett. 69B*, 309 (1977).
7. E. S. Fradkin and T. E. Fradkina, *Phys. Lett. 72B*, 343 (1978).
8. E. S. Fradkin and M. A. Vasiliev, *Phys. Lett. 72B*, 40 (1977).
9. T. Kugo and I. Ojima, *Phys. Lett. 73B*, 459 (1978); *Suppl. Prog. Theor. Phys. 66*, 1 (1979).
10. G. Curci and R. Ferrari, *Nuovo Cimento 35A*, 273 (1976).
11. M. Henneaux, *Phys. Rep. 126*, 1 (1985).
12. M. Henneaux, Forme hamiltonniene de l'integrale de chemin pour les théories à invariance de jauge, Thèse d'Agrégation, Bruxelles (1984); *Phys. Rev. Lett. 55*, 769 (1985); *Bull. Cl. Sci. Acad. R. Belg. 5e sér. LXXI*, 198 (1985).
13. R. Marnelius, *Phys. Lett. 99B*, 467 (1981); *Acta Phys. Polon. B 13*, 669 (1982).
14. P. A. M. Dirac, *Lectures on Quantum Mechanics*, Yeshiva University, Academic Press, New York, 1965; A. J. Hanson, T. Regge, and C. Teitelboim, *Constrained Hamiltonian Systems*, Accad. Naz. dei Lincei, Rome, 1976.

15. M. Henneaux and C. Teitelboim, B.R.S. quantization of generalized magnetic poles, to appear in E. S. Fradkin Festschrift, Adam Hilger, Bristol.
16. I. A. Batalin and G. A. Vilkovisky, *J. Math. Phys. 26,* 172 (1985).
17. I. A. Batalin and E. S. Fradkin, *Phys. Lett. 128B,* 303 (1983).
18. M. Henneaux, *Phys. Lett. 120B,* 179 (1983).
19. I. A. Batalin and E. S. Fradkin, *J. Math. Phys. 25,* 2426 (1984).
20. J. A. Shouten and W. v.d. Kulk, *Pfaff's Problem and its Generalizations,* Clarendon Press, Oxford, 1949, Chaps. VI and VII; see also S. Shanmugadhasan, *J. Math. Phys. 14,* 677 (1973).
21. B. L. Voronov and I. V. Tyutin, *Theor. Math. Phys. USSR 50,* 218 (1982).
22. P. G. Bergmann, *Rev. Mod. Phys. 33,* 510 (1961); L. D. Fadeev, *Theor. Math. Phys. 1,* 3 (1969).
23. M. Henneaux and C. Teitelboim, *Ann. Phys. (N.Y.) 143,* 127 (1982).

Chapter 11

Update on Anomalous Theories

R. Jackiw

1. WHAT IS AN ANOMALOUS THEORY?

It is known that the quantization procedure can spoil classical symmetries. The problem afflicts continuous chiral symmetries and gravitational symmetries of massless (Weyl) fermions, the former in any even-dimensional space-time, the latter in space-times with dimensionality $4k+2$, $k = 0, 1, \dots$. [Similar quantum breaking afflicts discrete symmetries (P, T) in odd dimensions, and scale/conformal symmetries in any dimension; we shall not be concerned with these.] As a consequence, the symmetry current, whose classical conservation is assured by Noether's theorem, ceases to be conserved after quantization. We call such a current *anomalous*; it possesses an *anomalous divergence*, and the coupling of gauge fields to this current becomes problematical [1].

The difficulty with gauge field couplings is seen from the equations of motion for the gauge field A_μ (taken to be an anti-Hermitian matrix in the Lie algebra of the internal symmetry group):

$$F_{\mu\nu} \equiv \partial_\mu A_\nu - \partial_\nu A_\mu + [A_\mu, A_\nu] \tag{1}$$

$$D_\mu F^{\mu\nu} \equiv \partial_\mu F^{\mu\nu} + [A_\mu, F^{\mu\nu}] = J^\nu \tag{2}$$

R. JACKIW • Center for Theoretical Physics, Massachusetts Institute of Technology, Cambridge, Massachusetts 02139.

Since $D_\mu D_\nu F^{\mu\nu} = 0$, Eq. (2) requires $D_\mu J^\mu = 0$. If, on the other hand, the current J^μ acquires an anomalous divergence,

$$D_\mu J^\mu = \text{anomaly} \neq 0 \tag{3}$$

it appears that the theory can be consistent only on the subspace where the anomaly vanishes. Moreover, aside from issues of dynamical consistency, questions about gauge invariance and unitarity arise. Finally, renormalizability of the quantum theory must be reexamined, since that desirable property is frequently linked with gauge invariance.

Another viewpoint is gotten from a functional integral formulation. Upon integrating over the chiral fermions that couple to the gauge field, one is left with an effective action:

$$W(A) = -i \ln \mathscr{D}(A) \tag{4}$$

$$\mathscr{D}(A) \equiv \det(\partial\!\!\!/ + A\!\!\!/) \tag{5}$$

Under a gauge transformation by the group element g,

$$A_\mu \to A_\mu^g \equiv g^{-1} A_\mu g + g^{-1} \partial_\mu g \tag{6}$$

which can be taken infinitesimally,

$$g = I + \theta + \cdots \tag{7}$$

$$\delta A_\mu = D_\mu \theta \tag{8}$$

$\mathscr{D}(A)$ changes as

$$\delta D \equiv \mathscr{D}(A + D\theta) - \mathscr{D}(A) = \int \operatorname{tr}(D_\mu \theta) \frac{\delta}{\delta A_\mu} \mathscr{D}(A) \tag{9}$$

The variation is the vacuum matrix element of the current in the presence of an external gauge field:

$$\langle J^\mu \rangle_A = -\frac{\delta}{\delta A_\mu} \mathscr{D}(A) \tag{10}$$

Hence

$$\delta \mathscr{D} = \int \operatorname{tr} \theta \mathscr{D}_\mu \langle J^\mu \rangle_A \tag{11}$$

and the effective action for chiral fermions is not gauge invariant when the fermion current is not conserved (in its matrix elements). Thus the total action, gauge field action $I(A)$ plus effective fermion action $W(A)$, loses gauge invariance and it is unclear whether the gauge invariant $I(A)$ can be consistently combined with the gauge noninvariant $W(A)$.

Models that possess the above-described gauge pathology are called *anomalous theories* [2].

An approach to anomalous theories that has been advocated in the past is to eliminate the anomaly, i.e., the (chiral) fermion content is adjusted so that the anomalous divergence vanishes [1]. More specifically, in four dimensions, the anomalous divergence reads [1]

$$(D_\mu J^\mu)_a \propto \frac{1}{24\pi^2} \operatorname{tr} T^a \partial_\mu \varepsilon^{\mu\alpha\beta\gamma}(A_\alpha \partial_\beta A_\gamma + \tfrac{1}{2} A_\alpha A_\beta A_\gamma) \tag{12}$$

Here, the T^a's comprise anti-Hermitian matrices that provide a basis for the Lie algebra, in the representation of the fermions:

$$A_\alpha = A_\alpha^a T^a \tag{13}$$

The proportionality constant is fixed by the number and chirality of fermions that interact with A_μ. The most important term in the divergence is the first. Because of cyclicity of the trace, it involves

$$\frac{i}{2} d_{abc} \equiv \operatorname{tr} T^a \{T^b, T^c\} \tag{14}$$

where the "b" and "c' indices refer to the gauge fields that couple to the current "a", as represented by the triangle diagram of Fig. 1.

The trace is unaffected by group transformations, hence an invariant statement for the absence of anomalies in four dimensions is that fermions belong to representations for which d_{abc} vanishes. (One can show that the rest of the anomaly also vanishes once d_{abc} does.)

This requirement of anomaly cancellation has been widely accepted and has two notable successes: First, when applied to the electroweak unified $SU(2) \times U(1)$ theory, which is potentially anomalous, it predicts that the number of quarks balances against the number of leptons, thus providing us with the only theoretical explanation of this apparently true experimental fact. Second, when applied to the construction of superstring models, the rank of the internal symmetry group is fixed uniquely at 16 and the group is essentially predicted. All the other imponderables in the string program make this one fixed fact very important indeed.

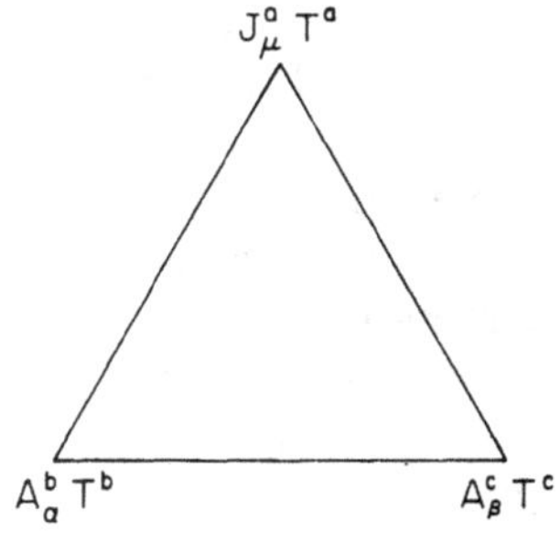

Figure 1. Triangle graph that spoils gauge invariance in a four-dimensional chiral gauge theory.

Precisely because of the central role that anomaly cancellation plays in modern theoretical particle physics, it is worthwhile looking more deeply into the matter, and inquiring whether a gauge theory that does possess anomalies is in fact meaningless, or whether some sense can be made of it. The question is especially appropriate at the present time, when as it happens a consistent mathematical framework for discussing anomalies has emerged.

The conclusion of several investigations is that unitarity and positivity need not be destroyed by the anomaly. Renormalizability of an anomalous theory remains unsettled.

DISCUSSION

C. TEITELBOIM: I would like to ask you something very naive: Is it possible to have anomalies in quantum mechanics, as opposed to quantum field theory?

R. JACKIW: Yes, several examples are given in the exercises to my Les Houches lectures, reprinted in Ref. 1. The same examples, and others, are also found in a paper by Elitzur *et al.* [*Nucl. Phys. B273*, 93 (1986)].

TEITELBOIM: So there can be a case where something is anomalously broken, without infinities.

JACKIW: Quantum mechanics does not suffer from field-theoretic ultraviolet infinities. However, there are infinities associated with "infinite-dimensional matrices," etc. Indeed, in the examples mentioned by me, one is evaluating determinants of differential operators.

2. MATHEMATICALLY COHERENT FRAME FOR ANOMALIES

When anomalous nonconservation of currents was first encountered, this unexpected phenomenon was associated with the ultraviolet divergences of perturbation theory [1]. Later, a Hamiltonian analysis of the effect related it to symmetry-breaking aspects of the filled "negative energy sea"—again a quantity that cannot be described in finite terms [1].

However, more recent discussion of the anomaly make use of various well-defined mathematical characteristic classes: Chern–Pontryagin, Chern–Simons, etc., and attention is drawn to the fact that various quantities of interest like the anomalous divergence of a chiral current, the anomalous response to gauge transformation of a fermion determinant, etc., are expressed in terms of these [1]. The mathematical connection has come to a sharper focus in the characterization of an anomalous gauge theory by the fact that commutators of gauge transformation generators are anomalous, and do not follow the Lie algebra of the gauge group [3].

It was appreciated already in the earliest days of research on anomalies that fermion charge densities $\rho_a \equiv J_a^0$ satisfy anomalous commutators [1]. Since ρ_a generates gauge transformations only on the fermion degrees of freedom, it is natural, in a mathematical/algebraic framework for the gauge theory, to consider the complete generator, G_a, given by ρ_a supplemented by $-(DE)_a$. [E_α^i is the non-Abelian electric field F_a^{i0}—it is the negative of the momentum conjugate to the canonical coordinate A_a^i.] G_a generates the complete gauge transformation, which modifies the gauge field degrees of freedom as well. It was conjectured [3] and then verified [4] that the anomaly in the commutator of the G_a's is also given by a mathematically determined quantity:

$$[G_a(\mathbf{r}), G_b(\mathbf{r}')] = if_{abc}G_c(\mathbf{r})\delta(\mathbf{r}-\mathbf{r}') \pm \frac{i}{24\pi^2}\varepsilon^{ijk}\,\mathrm{tr}\{T^a, T^b\}\,\partial_i A_j\,\partial_k\delta(\mathbf{r}-\mathbf{r}') \tag{15}$$

The sign is determined by the fermion chirality, and we observe that d_{abc} controls the anomaly in the commutator as well. (The third representation matrix making up d_{abc} resides in the gauge field: $A_j \equiv A_j^c T^c$.)

A finite gauge transformation with $g = e^{\theta^a T^a}$ is implemented in the quantum theory by the unitary operator

$$U(g) = \exp\left(i\int \theta^a G_a\right) \tag{16}$$

The fact (15) that the Lie algebra of the generators does not follow that of the group, but possesses an extension, means that the composition law for the operators U differs from the group composition law:

$$g_1 g_2 = g_{12} \tag{17}$$

$$U(g_1)U(g_2) = e^{-i2\pi\omega_2(A;g_1,g_2)}U(g_{12}) \tag{18}$$

The additional phase, called a 2-cocycle, when expanded for g_1 and g_2 near the identity reproduces the extension in (15).

Thus, the anomaly phenomenon is just the statement that in the quantized field theory, the gauge group is represented projectively, as in (18).

Before proceeding with the discussion of anomalous theories, let us pause for a mathematical message about cocycles.

3. MATHEMATICAL ASIDE ON COCYCLES

The mathematical theory for representing transformation groups by unitary operators on a linear vector or Hilbert space meets its objective correlative—its specific application—in quantum mechanics, which naturally is concerned with operators acting on physical states. When one is dealing with an abstract group of transformations, g, which obey the composition law (17) and which transform variables q according to a definite rule,

$$g \underset{g}{\rightarrow} q^g \tag{19}$$

then the simplest representation of this on functions of q, $\Psi(q)$, is obtained by associating with g an operator $U(g)$, which implements (19) as,

$$U(g)qU^{-1}(g) = q^g \tag{20}$$

acts on $\Psi(q)$ according to,

$$U(g)\Psi(q) = \Psi(q^g) \tag{21}$$

and satisfies a composition law that parallels (17):

$$U(g_1)U(g_2) = U(g_{12}) \tag{22}$$

However, it is possible to elaborate this simplest realization of the transformation group by introducing phases into formulas (21) and (22). Such phases are called *cocycles* and the various conditions that they must satisfy have been elaborated in mathematics.

It has been known for some time that cocycles are present in the quantum mechanics of point particles. As mentioned above, modern gauge field theory makes use of them as well, to give a coherent mathematical framework for the anomaly phenomena.

3.1. 1-Cocycle

In the simplest generalization of (21) and (22), a phase factor is inserted in the action of the representation:

$$U(g)\Psi(q) = e^{-i2\pi\omega_1(q;g)}\Psi(q^g) \tag{23}$$

Consistency with (22) requires that ω_1 satisfy

$$\omega_1(q^{g_1}; g_2) - \omega_1(q; g_{12}) + \omega_1(q; g_1) = 0 \qquad \text{(mod integer)} \tag{24}$$

A quantity $\omega_1(q; g)$ depending on one member of the transformation group, g, and possibly on the variable acted upon, q, is called a *1-cocycle* if it obeys (24).

Representing Galilean boosts in quantum mechanics makes use of a 1-cocycle. The Abelian group of Galileo transformations of the position vector $\mathbf{r}$ (q in the general discussion) is defined to act as $\mathbf{r} \rightarrow \mathbf{r} + \mathbf{v}t$. The transformation is labeled by $\mathbf{v}$ (corresponding to g in the general discussion) and is implemented by the operator

$$U(\mathbf{v}) = e^{i\mathbf{v}\cdot(\mathbf{p}t - m\mathbf{r})} \tag{25}$$

[The second term in the exponential involving $m\mathbf{r}$ is present so that in the Galilean invariant theory of a nonrelativistic free particle with mass m, the exponent is the appropriate constant of motion: $(d/dt)(\mathbf{p}t - m\mathbf{r}) = \dot{\mathbf{p}}t + \mathbf{p} - m\dot{\mathbf{r}} = 0$ when $\mathbf{p} = m\dot{\mathbf{r}}$, $\dot{\mathbf{p}} = 0$.] One verifies with the help of the Heisenberg algebra

$$\begin{aligned} [r^i, r^j] &= 0 = [p^i, p^j] \\ [r^i, p^j] &= i\delta^{ij} \end{aligned} \tag{26}$$

that

$$\begin{aligned} U(\mathbf{v})\mathbf{r}U^{-1}(\mathbf{v}) &= \mathbf{r} + \mathbf{v}t \\ U(\mathbf{v})\mathbf{p}U^{-1}(\mathbf{v}) &= \mathbf{p} + m\mathbf{v} \end{aligned} \tag{27}$$

and that (22) is satisfied, but $U(\mathbf{v})$ acts on wave functions $\Psi(\mathbf{r})$ as in (22) with

$$2\pi\omega_1(\mathbf{r}; \mathbf{v}) = m\mathbf{v}\cdot\mathbf{r} + \tfrac{1}{2}mv^2t \tag{28}$$

which satisfies (24).

3.2. 2-Cocycle

In the next generalization, a phase is introduced into the composition law (22):

$$U(g_1)U(g_2) = e^{-i2\pi\omega_2(q; g_1, g_2)}U(g_{12}) \tag{29}$$

A consistency condition on ω_2 follows from the assumed associativity of the composition law. If

$$[U(g_1)U(g_2)]U(g_3) = U(g_1)[U(g_2)U(g_3)] \tag{30}$$

then

$$\omega_2(q^{g_1}; g_2, g_3) - \omega_2(q; g_{12}, g_3) + \omega_2(q; g_1, g_{23}) - \omega_2(q; g_1, g_2) = 0 \quad \text{(mod integer)} \tag{31}$$

When a quantity depends on two group elements, g_1 and g_2, and possibly on q, and also satisfies (31), it is called a *2-cocycle.* Representations that make use of 2-cocycles are called *projective* or *ray* representations and they occur frequently in quantum mechanics.

Indeed, the Heisenberg commutator algebra (26) indicates that translations on phase space: $\mathbf{r} \to \mathbf{r} + \mathbf{a}$, $\mathbf{p} \to \mathbf{p} + \mathbf{b}$ are represented by the operator

$$U(\mathbf{a}, \mathbf{b}) = e^{i(\mathbf{a}\cdot\mathbf{p} - \mathbf{b}\cdot\mathbf{r})} \tag{32}$$

i.e.,

$$U(\mathbf{a},\mathbf{b})\mathbf{r}U^{-1}(\mathbf{a},\mathbf{b}) = \mathbf{r}+\mathbf{a}$$
$$U(\mathbf{a},\mathbf{b})\mathbf{p}U^{-1}(\mathbf{a},\mathbf{b}) = \mathbf{p}+\mathbf{b} \tag{33}$$

$U(\mathbf{a}, \mathbf{b})$ composes according to (29) with

$$2\pi\omega_2(\mathbf{r};\, \mathbf{a}_1\mathbf{b}_1,\, \mathbf{a}_2\mathbf{b}_2) = \tfrac{1}{2}(\mathbf{a}_1\cdot\mathbf{b}_2 - \mathbf{a}_2\cdot\mathbf{b}_1) \tag{34}$$

The U's associate and ω_2 obeys the consistency condition (31). (Coordinate and momentum shifts form a symmetry operation for a free particle: Galilean transformations supplemented by spatial translations are realized in this way.) So, following Weyl and Bargmann, one may accurately say that the essence of quantum mechanics is the 2-cocycle, leading to ray or projective representation of the Abelian translation group.

When dealing with a continuous or Lie group of transformations, the discussion may be carried out in infinitesimal terms. Corresponding to the finite group element g, there is the infinitesimal quantity θ, $g = e^{\theta}$, and the composition law (17) is reflected in a Lie algebra:

$$[\theta_1, \theta_2] = \theta_{12} \tag{35}$$

Suppose further that we are dealing with dynamical variables q, governed by Lagrangian L. The infinitesimal action of the transformation on q is given by an infinitesimal version of (19).

$$q \to q + \delta_\theta q \tag{36}$$

When no cocycles occur, the representative of θ, that is, the infinitesimal generator G, where $U(g) = e^{iG}$, is given by $(\delta_\theta q)\partial L/\partial \dot{q} = (\delta_\theta q)p$, so that its action on wave functions is realized by $(\delta_\theta q)(1/i)\partial/\partial q$. Moreover, the generators satisfy the Lie algebra of the group (35),

$$[iG_1, iG_2] = iG_{12} \tag{37}$$

which is a consequence of (17) for infinitesimal quantities.

In the presence of a 1-cocycle, the generator becomes

$$G = (\delta_\theta q)p - \delta\omega_1 \tag{38}$$

where $\delta\omega_1$ is (proportional to) the infinitesimal part of the 1-cocycle; compare with (25) and (28). The 1-cocycle condition (24), in infinitesimal form, ensures that the modified generators (38) continue to satisfy the Lie algebra of the group (35).

A 2-cocycle indicates that the generators' Lie algebra acquires an extension:

$$[iG_1, iG_2] = iG_{12} - \delta\omega_2 \tag{39}$$

Here $\delta\omega_2$ is [proportional to] the infinitesimal portion of the 2-cocycle, whose consistency condition (31) ensures that (39) does not contradict the Jacobi identity,

$$[G_1,[G_2, G_3]]+[G_2,[G_3, G_1]]+[G_3,[G_1, G_2]]=0 \tag{40}$$

This is the infinitesimal version of (30). The nonvanishing Heisenberg commutator (26) is an example of an extended Lie algebra (39).

3.3. 3-Cocycle

Not unexpectedly, the 3-cocycle involves abandoning associativity (30), and for infinitesimal generators, the Jacobi identity (40) fails. It is important to appreciate that nonassociating quantities cannot be represented by well-defined linear operators acting on a vector or Hilbert space since, by definition, operations on vectors necessarily associate. Hence, the quantities discussed below are abstract, algebraic objects obeying formal relations.

To encounter a 3-cocycle, we replace (30) by a formula that includes a phase:

$$[U(g_1)U(g_2)]U(g_3) = e^{-i2\pi\omega_3(q;g_1,g_2,g_3)}U(g_1)[U(g_2)U(g_3)] \tag{41}$$

By considering fourfold products, and associating in different ways, it is established that consistency requires that ω_3 satisfy the *3-cocycle* condition:

$$\begin{aligned}&\omega_3(q^{g_1};\, g_2, g_3, g_4) - \omega_3(q;\, g_{12}, g_3, g_4) + \omega_3(q;\, g_1, g_{23}, g_4)\\ &\quad -\omega_3(q;\, g_1, g_2, g_{34}) + \omega_3(q;\, g_1, g_2, g_3) = 0 \qquad (\text{mod integer})\end{aligned} \tag{42}$$

This arcane structure in fact arises in elementary Hamiltonian dynamics. Consider a particle with mass m and charge e, moving at $\mathbf{r}$ in an external magnetic field $\mathbf{B}(\mathbf{r})$. The dynamical equations express the Lorentz force law:

$$\begin{aligned} m\dot{\mathbf{v}} &= e\mathbf{v}\times\mathbf{B}\\ \dot{\mathbf{r}} &= \mathbf{v}\end{aligned} \tag{43}$$

The velocity of light is scaled to unity. We do not assume that $\mathbf{B}$ is necessarily divergence free, i.e., there may be magnetic sources. Since a magnetic field does no work, the energy is purely kinetic, and the Hamiltonian does not see $\mathbf{B}$:

$$H = \tfrac{1}{2}mv^2 \tag{44}$$

In order that the dynamical equations (43) emerge as canonical Hamiltonian equations, we postulate the following commutation relations for $\mathbf{r}$ and $\mathbf{v}$ (in the classical theory, these would be Poisson brackets):

$$[r^i, r^j] = 0 \tag{45a}$$

$$[r^i, mv^j] = i\delta^{ij} \tag{45b}$$

$$[mv^i, mv^j] = ie\varepsilon^{ijk}B^k \tag{45c}$$

Thus, the magnetic field, which is invisible in (44), reappears in (45c) so that (43) may be regained as

$$\begin{aligned} \dot{r} &= i[H, \mathbf{r}] = \mathbf{v} \\ m\dot{\mathbf{v}} &= i[H, m\mathbf{v}] = e\mathbf{v} \times \mathbf{B} \end{aligned} \tag{46}$$

If now (45) is subjected to the test of the Jacobi identity, we find that triple velocity commutators sum to

$$[mv^1, [mv^2, mv^3]] + [mv^2, [mv^3, mv^1]] + [mv^3, [mv^1, mv^2]] = e\boldsymbol{\nabla} \cdot \mathbf{B} \tag{47}$$

When $\boldsymbol{\nabla} \cdot \mathbf{B}$ is zero, we are dealing with an associative Lie algebra, and $m\mathbf{v}$ may be represented by $(1/i)\boldsymbol{\nabla} - e\mathbf{A}, \mathbf{B} = \boldsymbol{\nabla} \times \mathbf{A}$. However, for $\boldsymbol{\nabla} \cdot \mathbf{B} \neq 0$ a 3-cocycle is present.

To understand the nonassociativity, let us consider the finite quantities

$$U(\mathbf{a}) = e^{i\mathbf{a} \cdot m\mathbf{v}} \tag{48}$$

These represent translations of $\mathbf{r}$, in that $U(\mathbf{a})\mathbf{r}U^{-1}(\mathbf{a}) = \mathbf{r} + \mathbf{a}$. However, the representation is not faithful, because from (45c) it follows that

$$U(\mathbf{a}_1)U(\mathbf{a}_2) = e^{-ie\Phi}U(\mathbf{a}_1 + \mathbf{a}_2) \tag{49}$$

where Φ is the flux through the triangle at $\mathbf{r}$ formed from $\mathbf{a}_1$ and $\mathbf{a}_2$; see Fig. 2. Moreover, by considering the triple product $U(\mathbf{a}_i)U(\mathbf{a}_2)U(\mathbf{a}_3)$, associated in the two different ways, one finds

$$[U(\mathbf{a}_1)U(\mathbf{a}_2)]U(\mathbf{a}_3) = e^{-i2\pi\omega_3}U(\mathbf{a}_1)[U(\mathbf{a}_2)U(\mathbf{a}_3)] \tag{50}$$

where the 3-cocycle $2\pi\omega_3$ is e times the flux out of the tetrahedron at $\mathbf{r}$ formed from $\mathbf{a}_1$, $\mathbf{a}_2$, and $\mathbf{a}_3$; see Fig. 3.

When $\boldsymbol{\nabla} \cdot \mathbf{B} = 0$, no flux emanates from a closed surface; the 3-cocycle vanishes and associativity is regained. When there are sources, $\boldsymbol{\nabla} \cdot \mathbf{B} \neq 0$, the flux is nonzero , but associativity will prevail if ω_3 is an integer, since then $e^{-i2\pi\omega_3} = 1$. This requirement forces (1) $\boldsymbol{\nabla} \cdot \mathbf{B}$ to consist of delta functions so that the total flux not vary continuously when the $\mathbf{a}_i$ change, i.e., sources must be monopoles; (2) since a monopole of strength g gives rise to flux $4\pi g$, hence, produces the cocycle $2eg$, eg must satisfy Dirac's quantization condition, i.e., it must be an integral multiple of 1/2. In this way, the removal of the 3-cocycle, which is necessary for conventional

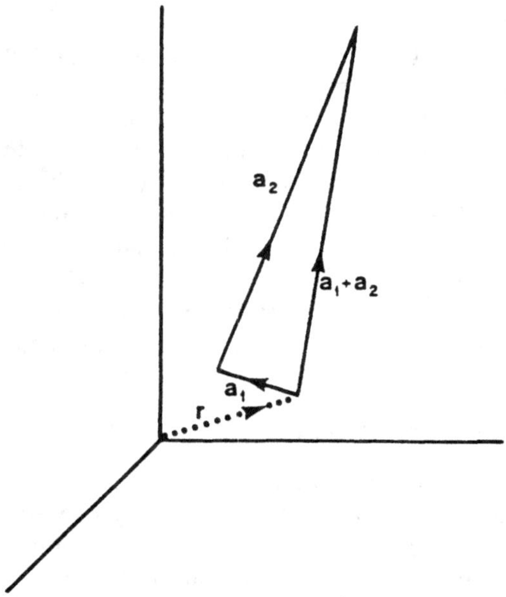

Figure 2. Triangle at r, defined by two translations a_1 and a_2, through which the flux Φ is calculated.

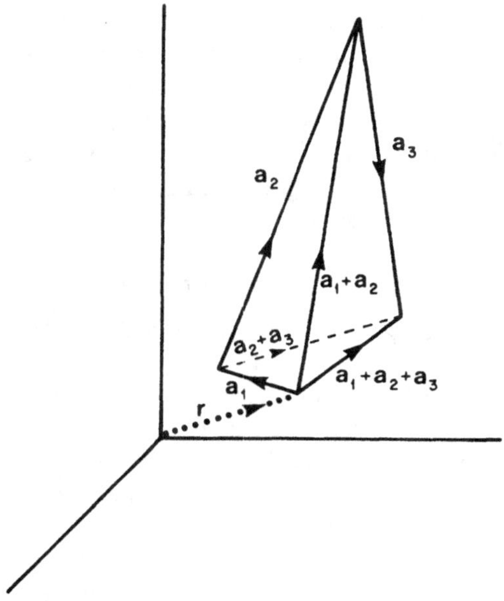

Figure 3. Tetrahedron at point r, defined by three translations a_1, a_2, and a_3. The 3-cocycle is proportional to the flux out of the tetrahedron.

quantum mechanics with associative operators on Hilbert space, limits magnetic sources to quantized Dirac monopoles. Other magnetic sources lead to a nonassociative algebra.

Finally, note that even for quantized monopoles, where the finite 3-cocycle is invisible since $e^{-i2\pi\omega_3} = 1$, there is a remnant of nonassociativity in the nonvanishing of the right-hand side of (47), which for a monopole at $\mathbf{r}_0$, reads $4\pi eg\delta(\mathbf{r}-\mathbf{r}_0) = 2\pi n\delta(\mathbf{r}-\mathbf{r}_0)$. This does not interfere with an operator formulation on a space with one point—the location of the monopole—excluded.

Further generalization is possible; one may define 4- and higher cocycles, but thus far no role has been found for these in physics.

3.4. Beyond Quantum Mechanics

The above concerns quantum mechanics. In quantum field theory, cocycles arise in the description of anomalies, and we see in (15) the role that a 2-cocycle plays. The 1-cocycle also is relevant: it measures the gauge noninvariance of the fermionic determinant (5). Define a quantity $\omega_1(A; g)$ by

$$\mathscr{D}(A^g) = e^{-2\pi i\omega_1(A;g)}\mathscr{D}(A) \tag{51}$$

From the gauge group composition law, $g_1 g_2 = g_{12}$, it follows that $\mathscr{D}((A^{g_1})^{g_2}) = \mathscr{D}(A^{g_{12}})$. Hence $\omega_1(A; g)$ must satisfy (24), i.e., it is a 1-cocycle. We further recognize that the haphazard-looking formula (12) must coincide with the infinitesimal, first-order in θ, contribution to $\omega_1(A; g)$, because according to (5) and (11) the anomalous divergence is an infinitesimal 1-cocycle. Moreover, from the 1-cocycle condition we see that the quantity $2\pi\omega_1(A; h)$ transforms under a gauge transformation $A \to A^g$, $h \to g^{-1}h$ exactly the same way as $W(A)$ does in (4):

$$\begin{aligned} W(A^g) &= W(A) - 2\pi\omega_1(A; g) \\ 2\pi\omega_1(A^g; g^{-1}h) &= 2\pi\omega_1(A; h) - 2\pi\omega_1(A; g) \end{aligned} \tag{52}$$

Therefore, one may view the 1-cocycle as an effective action that transforms under gauge transformations in the same anomalous way as the effective action of chiral fermions. Necessarily there is present a field other than the gauge field A, viz., the "chiral" field h. When the 1-cocycle is used in this way it is called the *Wess-Zumino term* [1].

Please note that in (51) and (52) the cocycle functions in a way somewhat different from the representation theory examples discussed earlier: unlike in (23), ω_1 goes not arise from the action of an operator on a vector.

What about field theoretic 3-cocycles? There is evidence of Jacobi identity violation for commutators of spatial components of fermionic currents even in a noninteracting theory [1, 5]. However, the structure of the commutators is ambiguous and fraught with divergences. Moreover, spatial components are involved, but they are not generators of symmetry transformations. Hence, no coherent algebraic framework has been given for this pathology.

Another commutator anomaly, hinting at an anomaly-related 3-cocycle, has been encountered in the course of verifying (15) [6]. When the commutator of two electric fields is computed in perturbation theory, one finds a nonvanishing result:

$$[E^i_a(\mathbf{r}), E^j_b(\mathbf{r}')] = \frac{i}{48\pi^2}\,\varepsilon^{ijk} d_{abc} A^k_c \delta(\mathbf{r}-\mathbf{r}') \tag{53}$$

If we view **E** as a "velocity" in field space—an interpretation that is appropriate since $\mathbf{E} = -\dot{\mathbf{A}}$ in the Weyl, $A^0 = 0$, gauge, then the right-hand side of (53) defines a $U(1)$ curvature in field space—a kind of functional "magnetic field"; compare (45c). (Commutators with **A**—a "coordinate" in field space—show no anomalies.) Moreover, the triple **E** commutator fails to satisfy the Jacobi identity.

Finally, we note that in string theory, violations of the Jacobi identity—potential 3-cocycles—have been found [7].

However, none of these fascinating structures have been coherently described by a nonassociative algebra. It is interesting to recall that already many years ago in a well-known body of work, P. Jordan investigated nonassociative algebras for quantum mechanics. Also, P. Dirac predicted a physical role for nonassociativity (1969 Rouse Ball lecture in Cambridge University, as communicated to the author by H. Gottlieb).

DISCUSSION

I. SCHMIDT: When you introduce a 3-cocycle, as in (50), do you still assume that a 2-cocycle arises in the composition, as in (29)?

R. JACKIW: In general, nothing is said about the composition law (29). But it is easy to show that if a phase is present there and also in (50), then (31), which is a consequence of associativity, is modified by the occurrence of the 3-cocycle on the right-hand side. Hence in that case the phase in composition law should not be called a cocycle.

M. HENNEAUX: Are the velocities which fail to satisfy the Jacobi identity linear operators?

JACKIW: No, they are not; linear operators acting on vectors necessarily associate by definition.

D. GROSS: Is it that $m\mathbf{v}$ cannot be $\mathbf{p} - e\mathbf{A}$ because $\mathbf{A}$ is not well defined?

JACKIW: $\mathbf{A}$ has not been introduced. I am giving a gauge-invariant discussion, without $\mathbf{A}$. Indeed, we cannot have a vector potential in the region that $\nabla \cdot \mathbf{B} \neq 0$.

GROSS: What goes wrong if you don't have associativity?

JACKIW: It is not clear. Certainly, you cannot have conventional quantum mechanics with linear operators on a Hilbert space. Perhaps you can do something with density matrices, but I don't know how far you can go.

HENNEAUX: But even when the Dirac quantization condition for a monopole holds, you cannot realize the velocities as linear operators.

JACKIW: Since the failure of the Jacobi identity for a monopole is located at one point—the position of the monopole—one can realize $m\mathbf{v}$ as $\mathbf{p} - e\mathbf{A}$ away from that point. However, as is well known, $\mathbf{A}$ is not globally defined.

P. VAN NIEUWENHUIZEN: If your formulas are not about linear operators, to what do they refer?

JACKIW: They are algebraic relations between quantities that I merely write on the blackboard. Whether they have some realization (representation) remains to be determined.

4. APPROACHES TO QUANTIZING AN ANOMALOUS THEORY

Since the gauge transformation generator $\mathbf{G}_a$ coincides with the time component of $-D_\mu F^{\mu\nu} + J^\nu$, the field equation motion (2) requires that $\mathbf{G}_a$ vanish—this is Gauss's law. Moreover, if the spatial components of the field equation vanish, so must the time component, as a consequence of Lorentz invariance.

In a Hamiltonian, classical canonical or quantum mechanical, context $\mathbf{G}_a = \rho_a - (\mathbf{D} \cdot \mathbf{E})_a$ involves only canonical variables and no time derivatives ($-\mathbf{E}_a$ is the canonical momentum in the Weyl gauge), hence its vanishing must be viewed as a constraint rather than an equation of motion.

Constraints are conveniently divided into first class, when they close under Poisson bracketing, and second class otherwise. For first-class constraints, the quantization procedures is standard, except that a further condition is imposed: physical states must be annihilated by the constraint. This is how a nonanomalous gauge theory is quantized, and the requirement $G_a|\Psi\rangle = 0$ ensures that states are invariant against homotopically trivial gauge transformations.

Second-class constraints cannot annihilate states, because their commutators do not close and a modification of the quantization rules is required to implement them.

Therefore, it is natural to suggest that an anomalous gauge theory can be consistently quantized, provided that Gauss's law is recognized as a second-class constraint and that the quantization procedure is appropriately altered [8]. Moreover, it has been conjectured that this procedure produces an effective action containing the Wess–Zumino term (1-cocycle), in addition to the gauge field and chiral fermion contributions [8]. In this way, the total action would regain gauge invariance: the anomalous response to gauge transformations of the chiral fermions would be canceled by the explicit gauge noninvariance of the Wess–Zumino term. The chiral field of the Wess–Zumino term signals the emergence of new degrees of freedom, which are seen in this scenario as a consequence of the anomaly.

This program of modified canonical quantization has thus far not been realized. The principal difficulty is that the second-class nature of the constraint algebra emerges as a quantization anomaly only after the theory has already been quantized by conventional methods.

Nevertheless, various indicators are encouraging.

4.1. Decoupling Anomaly-Canceling Fermions

Consider a potentially anomalous theory with spontaneous gauge symmetry breaking, effected by the Higgs mechanism, and fermion content properly adjusted to cancel the anomaly. By setting various parameters in the Higgs sector to infinity one may decouple some of the chiral fermions, and for the remaining ones the anomaly no longer cancels. It is found that the decoupling fermions leave behind the Wess–Zumino term, so that gauge invariance is retained. (The chiral field here is the Higgs field [9].)

The above may be viewed as a method for quantizing an anomalous gauge theory. Indeed, there is precedent for effecting difficult canonical quantization by first enlarging the theory in such a way that quantization is straightforward, and then removing the addition by sending parameters to infinity [10]. Note, however, that here the chiral field (Higgs field) retains a kinetic term, which is not needed for gauge invariance and may upset renormalizability.

4.2. Integrating over All Gauge Potentials

It has been observed that an effective action for chiral fermions that includes the Wess–Zumino term can be obtained from the functional integral, supplemented by the Faddeev–Popov trick. One begins with a formal functional integral over *all* configurations of the gauge potential, including gauge copies [11]:

$$Z = \int \mathscr{D}A \exp i[I(A) + W(A)] \tag{54}$$

[The chiral fermion fields have been integrated to yield $W(A)$.] Next unity is resolved in the Faddeev–Popov way:

$$1 = \Delta_f(A) \int \mathscr{D}g\,\delta[f(A^{g^{-1}})] \tag{55}$$

Here $f(A) = 0$ is a gauge fixing, and $\Delta_f(A)$ is the gauge invariant Faddeev–Popov determinant which renders (55) true. Upon multiplying (54) by unity, as presented in (55), changing orders of functional integration, and changing the integration variable from A to A^g, one is left with the announced results:

$$\begin{aligned} Z &= \int \mathscr{D}A\mathscr{D}_g\Delta_f(A)\delta[f(A)] \exp i[I(A^g) + W(A^g)] \\ &= \int \mathscr{D}A\mathscr{D}g\Delta_f(A)\delta[f(A)] \exp i[I(A) + W(A) - 2\pi\omega_1(A^g)] \end{aligned} \tag{56}$$

However, one cannot view this as a true derivation from quantum mechanical first principles. The Faddeev–Popov trick is just that—a trick that short-circuits a lengthy canonical analysis of constraints. It is known to be a valid trick in conventional gauge theories, but misses the mark in more subtly constrained situations. In the present context it is not at all clear why one should start, as in (54), with an integral over *all* gauge potentials. One may worry that integrating over all gauge configurations will destroy positivity and/or unitarity, owing to the indefinite metric. Indeed, a soluble example, which we now discuss, shows that unitarity can be lost in this way.

5. CHIRAL SCHWINGER MODEL

In view of the uncertainties mentioned above with the general program, it is very satisfying that in some simple models one can analyze the situation completely [12]. Especially useful is the chiral Schwinger model: two-dimensional electrodynamics with massless Weyl fermions of one chirality interacting with $U(1)$ gauge field. The Lagrangian is

$$\begin{aligned} \mathscr{L} &= -\frac{1}{4} F^{\mu\nu}F_{\mu\nu} + \bar{\psi}\gamma^\mu\left[i\partial_\mu + eA_\mu \frac{(1+i\gamma_5)}{2}\right]\psi \\ \gamma_5 &= i\gamma^0\gamma^1 \end{aligned} \tag{57}$$

Owing to the well-known simplifications that hold in two dimensions, the chiral fermion determinant may be evaluated. The effective action is

$$I_{\text{eff}} = I(A) + \frac{e^2}{8\pi}\int A_\mu\left[ag^{\mu\nu} - (g^{\mu\alpha} + \varepsilon^{\mu\alpha})\frac{\partial_\alpha\,\partial_\beta}{\Box}(g^{\beta\nu} - \varepsilon^{\beta\nu})\right]A_\nu \tag{58}$$

where a is a constant, left undetermined in the evaluation of the determinant. For no value of a is the answer gauge invariant—that is why the theory is anomalous. Since a cannot be determined, the strength of the contact term $\int A^2$ is arbitrary. In a gauge invariant theory contact terms are fixed by the requirement of gauge invariance; here we have no such principle.

An equivalent description is in terms of a bosonized action,

$$I(A, \phi) = I(A) + \frac{1}{2}\int d^2x \times \left[\partial_\mu\phi\partial^\mu\phi + \frac{e}{\sqrt{\pi}}(g^{\mu\nu} - \varepsilon^{\mu\nu})\partial_\mu\phi A_\nu + \frac{ae^2}{4\pi}A_\mu A^\mu\right] \tag{59}$$

where the a term now arises from ambiguities in the bosonization procedure. Functionally integrating over ϕ in (59) yields (58).

The theory described by (58) or (59) may be solved. One finds for $a \geq 1$ a unitary, positive definite model. The vector meson acquires mass m:

$$m^2 = \frac{e^2}{4\pi}\frac{a^2}{a-1} \tag{60}$$

This lies anywhere between e^2/π and infinity, in contrast to the conventional Schwinger model, where gauge invariance fixes the corresponding constant a and the gauge boson mass.

It is noteworthy that a consistent, but not completely determined, theory has emerged. (The one-parameter ambiguity at first caused confusion [13], but now has been rederived by many different methods [14].) In a sense, one can view the above calculation, with its loss of gauge invariance, as a calculation within the gauge invariant chiral action with Wess–Zumino term (56), in the "unitary" gauge $g = 1$, where the Wess–Zumino term vanishes. Alternatively, one can regard the result as coming from an anomaly-free theory, where additional anomaly-canceling fermions have been decoupled, by sending parameters to infinity. However, neither of these alternative viewpoints explains why only a range of the undetermined parameter [$a \geq 1$] gives a unitary theory, while unitarity is lost outside this range [$a < 1$].

The bosonized action explicitly incorporates the anomaly, and canonically quantizing (59) does indeed involve second-class constraints. Carrying out the details of this quantization reproduces the results obtained by the action integral [15].

In all of these derivations, one fact remains: a consistent, Lorentz invariant [16] theory of a massive vector meson emerges from an anomalous gauge theory. Similar results can be established for the non-Abelian generalization, though the analysis is more formal and less explicit, since the model cannot be solved completely [17].

Unfortunately nothing is learned from the two-dimensional exercises about renormalizability; these two-dimensional theories are finite. Thus it is not known whether four-dimensional anomalous theories are renormalizable, but one may expect that anomalies do not spoil unitarity and positivity. This expectation is borne out by a formal canonical analysis of a bosonized action that describes the low-energy dynamics of an anomalous four-dimensional gauge theory [18], and by a formal BRST analysis of the complete gauge field plus fermionic action [19]. In *these* investigations the gauge symmetry is broken, and the gauge field is massive.

DISCUSSION

D. GROSS: How can you see that the constraint is second class before you quantize?

R. JACKIW: As I mentioned, that is precisely the obstacle to carrying out the constrained quantization program with anomalies—they arise only after the theory has been quantized, but then they change the way that the theory should be quantized.

C. TEITELBOIM: To get started you have to put in $\hbar$ classically.

JACKIW: Something like that. You can write an effective (bosonized) action, as in (59), which reproduces the anomalies by explicit (classical) reasoning. Then you can quantize, and you find you are dealing with second-class constraints.

6. GRAVITATIONAL THEORIES

In 2 mod 4 dimensions, chiral fermions induce gravitational anomalies [1]. This is to be expected since gravity theory may be formulated as a gauge theory, invariant against local Lorentz transformations. The current which now is the symmetric energy-momentum tensor $\theta_{\mu\nu}$, is not (covariantly) conserved in the presence of chiral fermions. Just as for gauge anomalies, which arise in 2 mod 2 dimensions, the two-dimensional case is the most important one, but gravity in two dimensions has its own pecularities, which have nothing to do with anomalies.

The usual equation governing gravity, involving the Einstein tensor, $G_{\mu\nu}$,

$$\begin{aligned} G_{\mu\nu} - \Lambda g_{\mu\nu} &= G\theta_{\mu\nu} \\ G_{\mu\nu} &\equiv R_{\mu\nu} - \tfrac{1}{2} g_{\mu\nu} g^{\alpha\beta} R_{\alpha\beta} \end{aligned} \tag{61}$$

where $R_{\mu\nu}$ is the Ricci curvature, Λ a cosmological constant, $g_{\mu\nu}$ the metric, and G the gravitational constant, requires conservation of $\theta_{\mu\nu}$, since the left-hand side possesses this property. However, in two dimensions (61) cannot be postulated, since $G_{\mu\nu}$ vanishes identically. Correspondingly, the quantity $\int d^2x\sqrt{-g}R$, $g \equiv \det g_{\mu\nu}$, $R \equiv g^{\mu\nu}R_{\mu\nu}$, whose variation gives $G_{\mu\nu}$ in dimension greater than 2, is a surface term in two dimensions, with vanishing variation.

Therefore, two-dimensional gravity must be based on an equation different from (61), and the anomaly problem for two-dimensional gravity becomes more subtle than a mere mismatch between the conservation properties of two sides of Einstein's equation, as in (61).

Nevertheless, the essence of the problem remains the same: the chiral fermion determinant (which may be computed exactly [20]) loses local Lorentz invariance (general coordinate invariance can always be maintained). Also, it possesses an arbitrariness, analogous to that parametrized by "a" in the gauge theory case.

The noninvariant chiral fermion effective action may be combined with an invariant action for two-dimensional gravity whose form has been derived from the three-dimensional Einstein-Hilbert action by dimensional reduction [21]:

$$I(g) = \int d^2x\, N\sqrt{-g}(R + 2\Lambda) \tag{62}$$

Here N is a scalar Lagrange multiplier. The analysis of the complete theory cannot be carried out completely, since the model is nontrivial. But again, formal indications support unitarity with massless excitations for a range of the arbitrary parameter [22].

7. CONCLUSIONS AND OUTLOOK

It should not be construed that my present pursuit of anomalous theories repudiates the anomaly cancellation principle that D. Gross and I enunciated in response to the Weinberg-Salam model in the 1970s. In its original formulation, that model is anomalous, yet a conventional quantization was proposed; this is inconsistent. The remedy we put forward cancels the anomaly with a quark-lepton balance, which is verified but not completely established experimentally: the top quark has not been discovered and the τ neutrino need not be different from the e and μ neutrinos. Of course, absence of the sixth quark can be accommodated at any given time if its mass is beyond experimental range. However, should ν_τ prove not to be a new particle, anomalies cannot cancel, and new ideas—perhaps those presented here—will be necessary.

Of course, renormalizability of a four-dimensional anomalous theory remains an open problem, which is under consideration [23]. The hope is that infinities arising from the anomaly cancel against those from the Wess–Zumino term. Verification of this is complicated in the absence of a kinetic term for the Wess–Zumino chiral field [24]. Should a positive result hold, it could very well be coupled with the emergence of a mass for the gauge fields, as is seen in all the examples studied thus far. Gauge symmetry breaking in electroweak theories would then appear in a new light: the weak forces are mediated by massive vector mesons whose mass breaks the gauge symmetry owing to chiral anomalies; these are absent in the chirally symmetric electromagnetic channel with its massless photon.

While uncertainties remain about four-dimensional anomalous theories, the complete success in two dimensions counts as an important result, in view of the many physical applications and the central role for the string program that two-dimensional models enjoy.

ACKNOWLEDGMENT. This work was supported in part by funds provided by the U.S. Department of Energy (D.O.E.) under contract No. DE-AC02-76ER03069.

REFERENCES AND NOTES

1. For a review and references to the original literature, see S. Treiman, R. Jackiw, B. Zumino, and E. Witten, *Current Algebra and Anomalies*, Princeton University Press, Princeton, New Jersey/World Scientific, Singapore, 1985; *Anomalies, Geometry, Topology* (W. Bardeen and A. White, eds.), World Scientific, Singapore, 1985.
2. For earlier review see R. Jackiw, in: *Proceedings of the Oregon Meeting* (R. Hwa, ed.), World Scientific, Singapore, 1986, p. 772; R. Rajaraman, Second Asia Pacific Physics Conference, Indian Institute of Science (Bangalore) (1986) preprint; C. Viallet, *Super Field Theories* (H. C. Lee, V. Elias, G. Kunstatter, R. B. Mann, and K. S. Viswanathan, eds.), Plenum, New York, 1987, p. 399.
3. L. Faddeev, *Phys. Lett. 145B*, 81 (1984); J. Mickelsson, *Commun. Math. Phys. 97*, 361 (1985); I. Singer, *Asterisque*, 323 (1985).
4. S.-G. Jo, *Phys. Lett. 163B*, 353 (1985); M. Kobayashi, K. Seo, and A. Sugamoto, *Nucl. Phys. B273*, 607 (1986). Let me note here that other purported verifications, using regulated fixed-time procedures for calculating the anomalous commutator, in contrast to the BJL method of the above-cited investigations, are in fact incomplete: L. Faddeev and S. Shatashvili's paper, *Phys. Lett. 167B*, 225 (1986), contains errors, while I. Frenkel and I. Singer did not complete their calculations. A non-BJL analysis, which apparently also confirms the conjectured form of the commutator, is by A. Niemi and G. Semenoff, *Phys. Rev. Lett. 56*, 1019 (1986). This interesting paper relates the anomaly phenomenon to Berry's adiabatic phase; see also G. Semenoff, *Super Field Theories* (H. C. Lee, V. Elias, G. Kunstatter, R. B. Mann, and K. S. Viswanathan, eds.), Plenum, New York, 1987, p. 407.
5. Recent investigations with references to the older literature are D. Levy, *Nucl. Phys. B282*, 367 (1987); Y.-Z. Zhang, *Phys. Lett. 189B*, 149 (1987).
6. Jo, Niemi and Semenoff, Ref. 4.
7. G. Horowitz and A. Strominger, *Phys. Lett. 185B*, 45 (1987); A. Strominger, *ibid. 187B*, 149 (1987).

8. L. Faddeev, in: *Supersymmetry and its Applications* (G. Gibbons, S. Hawking, and P. Townsend, eds.), Cambridge University Press, Cambridge, England, 1985, p. 41; Faddeev and Shatashvili, Ref. 4.
9. E. D. 'Hoker and E. Farhi, *Nucl. Phys. B248*, 59, 77 (1984).
10. J. Reiff and M. Veltman, *Nucl. Phys. B13*, 545 (1969).
11. O. Babelon, F. Schaposnik, and C. Viallet, *Phys. Lett. 177B*, 385 (1986); A. Kulikov, Setpukhov preprint No. IHEP 86-83 (1986); K. Harada and I. Tsutsui, *Phys. Lett. 183B*, 311 (1987); N. Falck and G. Kramer, *Ann. Phys. (N.Y.) 176*, 330 (1987).
12. R. Jackiw and R. Rajaraman, *Phys. Rev. Lett. 54*, 1219, 2060(E), *55*, 2224(C) (1985).
13. C. Hagen, *Phys. Rev. Lett. 55*, 2223(C) (1985); A. Das, *Phys. Rev. Lett. 55*, 2126 (1985); H. Ito, Nagoya University preprint (1986).
14. R. Banerjee, *Phys. Rev. Lett. 56*, 1889 (1986); J. Webb, *Z. Phys. C31*, 301 (1986). M. Chanowitz, *Phys. Lett. 171B*, 280 (1986); K. Harada, T. Kubota, and I. Tsutsui, *Phys. Lett. 173B*, 77 (1986); I. Halliday, E. Rabinovici, A. Schwimmer, and M. Chanowitz, *Nucl. Phys. B268*, 413 (1986); H. Girotti, H. Rothe, and K. Rothe, *Phys. Rev. D34*, 592 (1986); F. Schaposnik and J. Webb, *Z. Phys. C 34*, 367 (1987); R. Ball, *Phys. Lett. 183B*, 315 (1987); K. Harada and I. Tsutsui, *Prog. Theor. Phys. 78*, 878 (1987).
15. R. Rajaraman, *Phys. Lett. 154B*, 305 (1985); J. Lott and R. Rajaraman, *Phys. Lett. 165B*, 321 (1985); H. Girotti, H. Rothe, and K. Rother, *Phys. Rev. D 33*, 514 (1986); K. Harada and I. Tsutsui, *Z. Phys. C* (to be published).
16. Lorentz noninvariant results have also been obtained by Halliday *et al.* and Chanowitz, Ref. 14; A. Niemi and G. Semenoff, *Phys. Lett. 175B*, 439 (1986). But these authors use a Lorentz noninvariant gauge, which cannot be justified in a gauge noninvariant theory.
17. R. Rajaraman, *Phys. Lett. 162B*, 148 (1985); Y.-Z. Zhang, Northwest University (Xian) preprint No. NWU-IMP-86-11 (1986).
18. R. Rajaraman, *Phys. Lett. 184B*, 369 (1987).
19. S. Rajeev, MIT preprint CTP No. 1405 (1986).
20. H. Leutwyler, *Phys. Lett. 153B*, 65 (1985); J. Lott, IHES (Bures) preprint No. IHES/P/86 (1986).
21. R. Jackiw, in: *Quantum Theory of Gravity* (S. Christensen, ed.) Hilger, Bristol, UK, 1984, p. 403; *Nucl. Phys. B252*, 343 (1985); C. Teitelboim, in: *Quantum Theory of Gravity* (S. Christensen, ed.), Hilger, Bristol, UK, 1984, p. 327.
22. K.-K. Li, *Phys. Rev. D 34*, 2292 (1986); F. Schaposnik and H. Vucetich, LaPlata preprint (1986).
23. L. Faddeev (private communication).
24. E. D'Hoker and E. Farhi, in unpublished research, have convinced themselves that their model (Ref. 9), with a kinetic term for the Wess-Zumino chiral field, is not renormalizable.

Chapter 12

The Relativistic String and Its Nonrelativistic Disguises

Fernando Lund

1. INTRODUCTION

Currently there is much interest in the theory of strings,* as they provide hope of unifying a quantized version of gravity with the other interactions. String theory goes back to the work of Nambu and Veneziano, and in this chapter we take a new look at the Nambu string in interaction with a multiple-valued field. Although the interaction is local, the action integral involves not only the world surface of the (closed) string but also the volume it bounds. This outlook turns out to be useful in two apparently unrelated fields; vortex motion in a compressible, inviscid, fluid, and dislocation dynamics in an elastic solid. The study of vortex dynamics is of particular importance to the understanding of turbulence, where vortex stretching appears to be a major mechanism in the transfer of energy from larger to smaller scales [1]. Fully developed turbulence remains largely an unsolved mystery in spite of the very impressive progress made in recent years in the study of the transition mechanisms leading from orderly to chaotic motion [2], and an assessment of the role of compressibility and sound emission is relevant as a matter of principle so as to get a handle on a possible

* See D. J. Gross's lectures, Chapter 9 in this volume.

FERNANDO LUND • Departamento de Física, Facultad de Ciencias Físicas y Matemáticas, Santiago, Chile.

diagnostic of turbulent behavior, as well as for very pedestrian reasons such as understanding noise generation from jet aircraft [3]. Dislocation dynamics might be regarded as the dynamics of a special type of elastic singularity. Extension of the methods reported here to other types of singularities such as cracks seems feasible, mainly because cracks may be regarded as continuous distributions of dislocations, and one might hope for an improved understanding of the dislocation–crack interaction at high speeds, a problem whose relevance to fracture has been particularly emphasized by Thomson [4].

2. RELATIVISTIC STRING COUPLED TO A MASSLESS SCALAR FIELD [5]

Consider a classical massless scalar field ϕ, living in four-dimensional space-time, which is allowed not to be single valued. It may have a discontinuity, say f, across a timelike three-dimensional submanifold Ω, with boundary $\partial\Omega$, as for a solid angle. Its gradient $\partial_\mu \phi$ is, however, continuous. If, in addition, it is imposed that ϕ vanish at spacelike infinity, we have a well-defined boundary value problem that can be solved with the help of the Green's function $G(y) = \theta(y_0)\delta(y^2)$, the (retarded) solution of

$$\partial_\mu \partial^\mu G(y - y') = \delta^{(4)}(y - y')$$

as follows: Consider the identity

$$\int d^4y' \frac{\partial \phi(y')}{\partial y'_\mu} \frac{\partial G(y - y')}{\partial y'^\mu} = \int d^4y' \frac{\partial \phi(y')}{\partial y'_\mu} \frac{\partial G(y - y')}{\partial y'^\mu}$$

and integrate by parts the derivatives of ϕ on the left-hand side, and those of G on the right-hand side. The latter vanish on account of the single valuedness of $\partial_\mu \phi$, the asymptotic behavior of G, and the equation of motion for ϕ. We are then left with

$$\int d^4y' \frac{\partial}{\partial y'_u} \left(\phi \frac{\partial G}{\partial y'^\mu} \right) - \int d^4y' \, \phi \partial_\mu \partial^\mu G = 0$$

Using Gauss' theorem on the first term and the fact that ϕ has a discontinuity f across Ω we get

$$\phi(y) = f \int_\Omega dV^\mu \, \partial_\mu G(y - x) \tag{1}$$

where $dV^\mu = (1/24)\varepsilon^{\mu\nu\lambda\rho} dx_\nu \wedge dx_\lambda \wedge dx_\rho$ is the volume induced in Ω by the flat metric of the ambient space-time. Equation (1) can be written as a four-dimensional convolution

$$\phi = G * \tilde{s}$$

of a Green's function G with a source

$$\tilde{s}(y) = f \int_{\Omega} dV^{\mu}\, \partial_{\mu}\delta^{(4)}(y - x)$$

which has support on the three-dimensional submanifold Ω, which in turn may be thought of as the world history of a membrane or drum, and equation (1) gives the scalar field generated by a drum whose evolution is given. Note now that, the field ϕ being massless, only its gradients appear in the physically relevant four-momentum. Computation of this gradient gives, for points not in Ω,

$$\partial_{\mu}\phi = G * s_{\mu} \tag{2}$$

with

$$s_{\mu}(y) = f \int_{\partial\Omega} dx^{\nu} \wedge dx^{\lambda} \varepsilon_{\mu\alpha\nu\lambda} \partial^{\alpha}\delta^{(4)}(y - x)$$

This is a convolution with a source whose support lies on the boundary $\partial\Omega$ of Ω, which may be regarded as the world history of a closed string. We are thus led to the conclusion that physical quantities associated with the scalar field ϕ are determined by the evolution of the string, with the quantity f playing the role of coupling constant. Moreover, the fields (2) are singular at the source. More than one string will generate a field to be found by superposition, and the following question arises: Is it possible to determine the evolution of a test string in a prescribed external field, say, in the same sense that the Lorentz force determines the response of test charged particle to a prescribed electromagnetic field? The answer is a qualified yes, as we now proceed to spell out. The string, generating a field ϕ, moves in a yet-to-be-found way in an external field Φ. The latter is unaffected by whatever the string does, which is of course the practical implementation of its being a "test" one. The singularity of ϕ is a result of our having a filamentary string, which, from a physical point of view, means that we do not wish, or are unable, to resolve what happens at very short wavelengths. Thus, the string is "really" endowed with a thin core, outside of which everything happens as has been described above, and about whose inside we know nothing. The strategy is now to take the action integral

$$S = \int d^4y\, \partial_{\mu}\tilde{\phi}\partial^{\mu}\tilde{\phi}$$

with $\tilde{\phi} = \phi + \Phi$ and where the region of integration *excludes* the world history of the thin core region around the world history of the string.

Thereby, S becomes a functional of that world history and the external field and those world histories extremizing S determine the evolution sought. We then have that the action is the sum of three terms:

$$S = S_s + S_m + S_e = \int \partial_\mu \phi\, \partial^\mu \phi + 2\int \partial_\mu \phi\, \partial^\mu \Phi + \int \partial_\mu \Phi\, \partial^\mu \Phi$$

an external (S_e), mixed (S_m), and self- (S_s) action. Consider S_m first:

$$S_m = 2\int d^4y\, \partial_\mu(\phi \partial^\mu \Phi)$$

since Φ is supposed to be a solution of the wave equation throughout space. The integration must exclude not only a thin tube around the string but also a thin slab sandwiching the membrane of which the string is the boundary, since ϕ is discontinuous across it. With appropriate boundary conditions for Φ we have

$$S_m = 2\int_{\text{slab}} dV_\mu \phi\, \partial^\mu \Phi + 2\int_{\text{tube}} dV_\mu \phi\, \partial^\mu \Phi \tag{3}$$

For a very thin tube, say of thickness $\varepsilon \to 0$, the second integral on the right-hand side of (3) vanishes because Φ is supposed to be well behaved, $dV_\mu \sim \varepsilon$ and, from (2), $\partial_\mu \phi \sim \varepsilon^{-1}$ so that at worst $\phi \sim \log \varepsilon$. Taking then the limit $\varepsilon \to 0$ and using that ϕ has a discontinuity f across Ω,

$$S_M = 2f\int_\Omega dV_\mu\, \partial^\mu \Phi$$

which at first sight seems unsatisfactory since it is a functional of the world history of the whole membrane rather than its stringy boundary. The variation δS_m is, however,

$$\delta S_m = \frac{f}{2}\int_{\partial\Omega} \varepsilon_{\nu\lambda\rho\alpha} \partial^\alpha \Phi\, \delta x^\rho\, dx^\nu \wedge dx^\lambda \tag{4}$$

It has support only on the boundary $\partial\Omega$: the string history as well as initial and final membrane configurations.

Take now S_s, which, with the same reasoning that was used with S_m, can be written

$$S_s = f\int_\Omega dV_\mu\, \partial^\mu \phi + \int_{\text{tube}} dV_\mu \phi\, \partial^\mu \phi + \int_\infty dV_\mu \phi\, \partial^\mu \phi \tag{5}$$

where ϕ and $\partial_\mu \phi$ are given by (1) and (2). The last integral on the right-hand side is carried over the boundary of the volume over which the action S is defined. Moreover, from (2) we see that $\partial_\mu \phi$ is a curl, say $\partial_\mu \phi = \varepsilon_{\mu\nu\lambda\rho}\, \partial^\nu A^{\lambda\rho}$. This means that, because of Stokes' theorem, the integral over Ω can be transformed into an integral over $\partial\Omega$. Strictly speaking, (5) is a function of the world history of the string, and its extrema will determine the latter's evolution. In general, however, such extrema will yield integrodifferential equations reflecting the highly nonlocal nature of the string self-interaction. Indeed, the motion of a string point will be affected by the radiation from all other points of the string. This radiation having taken place previously, it will be necessary to specify not only position and a few derivatives at one time in order to determine the subsequent evolution, but the whole previous string history. Also, the contribution to S_s from the integral over the thin tube surrounding the dislocation depends on the detailed shape of the tube, something that is typical of radiation reaction forces from singularities [6]. Finally, the integral over Ω is divergent when $\varepsilon \to 0$, and this difficulty offers a way to look at the situation in a much simpler fashion: assume that the response of the test string to an external field is determined by the most divergent term in S_s. This is a decent approximation as long as string accelerations and curvatures are not very large. Neglecting then all terms that are finite when $\varepsilon \to 0$, we have

$$S_s \approx \frac{1}{4} f^2 \int_{\partial\Omega} dx^\alpha \wedge dx^\beta \int_{\partial\Omega'} dy_\alpha \wedge dy_\beta G(x-y)$$

where $y^\mu(\sigma', \tau') = x^\mu(\sigma', \tau') + \varepsilon^\mu$. Now we want to isolate the most divergent part of S_s, which comes from points $y \sim x$. This is a local approximation in which the effect of radiation from remote points of the string is neglected. Writing, for $\sigma \sim \sigma'$ and $\tau \sim \tau'$,

$$x^\mu(\sigma, \tau) - y^\mu(\sigma', \tau') \approx -\dot{x}^\mu \Delta\tau - x'^\mu \Delta\sigma - \varepsilon^\mu$$

with $\Delta\tau = \tau - \tau'$ and $\Delta\sigma = \sigma - \sigma'$ and carrying out the integrals over $\Delta\tau$, $\Delta\sigma$ with $G(x) = (1/4\pi)\theta(x^0)\delta(x^2)$ one obtains

$$S_s \simeq -\frac{f^2}{8\pi} \ln\frac{\delta}{\varepsilon} \int d\sigma\, d\tau \sqrt{-g} + \text{finite terms} \tag{6}$$

where $-g = (\dot{x} \cdot x')^2 - \dot{x}^2 x'^2$ is the determinant of the metric of the world-sheet of the string and the Nambu action* is recognized. Thus, in the local approximation that we have adopted the string responds to an external field as if it were endowed with a tension originating entirely in its own massless field. The evolution is determined by $\delta S_m + \delta S_s = 0$ where one must fix initially and finally not only the string variables x^μ but also some membrane

* See Ref. 7. The Nambu string in interaction with an antisymmetric tensor field was also considered by Cremmer and Scherk [8], Kalb and Ramond [9], and Lund and Regge [10].

of which it is a boundary. The parameter δ appearing in (6) is a cutoff needed to implement the local approximation. As is apparent, only the ratio δ/ε is relevant, and it is understood that $\delta/\varepsilon \to \infty$.

3. VORTEX FILAMENTS IN A SLIGHTLY COMPRESSIBLE FLUID [11]

Take the massless field ϕ of the previous section and write its wave equation with space and time separated out, and with dimensional units:

$$\frac{\partial^2 \phi}{\partial t^2} - c^2 \nabla^2 \phi = 0$$

If ϕ is interpreted as a velocity potential and c is a constant, this is the equation for sound waves in a fluid, from which particle velocity $\mathbf{v} = \nabla \phi$ and pressure $p = \partial \phi / \partial t$ are obtained. If $\phi(\mathbf{x}, t)$ is allowed to be multiple valued, with a discontinuity f when crossing a surface S with boundary B, the velocity $\mathbf{v}$, in spite of being locally a gradient, will have a nonvanishing curl $\nabla \wedge \mathbf{v} = \boldsymbol{\omega}$ (vorticity) with support on the curve B. This curve then is a vortex filament with circulation f, and equations (2) provide expressions for the sound generated by a vortex filament undergoing some prescribed motion. Rewriting equations (2) in time and space components yields

$$\nabla \phi(\mathbf{x}, t) = f \int dt' \, d\sigma \left(\nabla G \wedge \frac{\partial \mathbf{X}}{\partial \sigma} \right) + \frac{f}{c^2} \int dt' \, d\sigma \frac{\partial G}{\partial t} \left(\frac{\partial \mathbf{X}}{\partial t'} \wedge \frac{\partial \mathbf{X}}{\partial \sigma} \right)$$

$$\frac{\partial \phi(\mathbf{x}, t)}{\partial t} = f \int dt' \, d\sigma \left(\frac{\partial \mathbf{X}}{\partial t'} \wedge \frac{\partial \mathbf{X}}{\partial \sigma} \right) \cdot \nabla G$$

where $G = G(\mathbf{x} - \mathbf{X};\, t - t')$ is the retarded Green's function for the wave equation and $\mathbf{X}(t', \sigma)$ is the vortex filament, a time-dependent closed curve in three-dimensional space whose points are labeled by a Lagrangean parameter σ. Formally letting $c \to \infty$ we recover the Biot–Savart law for the velocity generated by a vortex filament in an incompressible fluid:

$$\mathbf{v}(\mathbf{x}, t) = \frac{f}{4\pi} \int d\sigma \frac{\partial \mathbf{X}}{\partial \sigma} \wedge \frac{(\mathbf{x} - \mathbf{X})}{|\mathbf{x} - \mathbf{X}|^3}$$

The flows generated by several filaments, including a continuous distribution thereof, are simply obtained by superposition. In order to find, next, the response of a vortex filament to an external flow in the incompressible limit one is tempted to simply use the reasoning of Section 2 in a space of one dimension less: since the equation obeyed by the velocity potential

$$\nabla^2 \phi = 0$$

is obtained by extremizing

$$\int dt\, d^3x (\nabla \phi)^2$$

the flow might be thought of as decomposed in a self-flow, generated by the vortex filament, and an external flow, and so on. There is a catch, however, in that the variational principle from which the equations of fluid dynamics follow is $\delta S = 0$ with

$$S = \int dt\, d^3x \left[\rho \frac{\partial \phi}{\partial t} + \frac{1}{2} \rho (\nabla \phi)^2 + \rho E \right] \tag{7}$$

where ρ is the density and E the internal energy. When the flow is adiabatic and incompressible,

$$S_0 = \rho_0 \int dt\, d^3x \left[\frac{\partial \phi}{\partial t} + \frac{1}{2} (\nabla \phi)^2 \right]$$

and the term involving $\partial\phi/\partial t$ may be omitted only when the volume of integration is time independent. This is not the case for us, who wish to exclude a thin tube surrounding the vortex filament, and it must be retained. On the other hand, in the incompressible approximation there is neither radiation nor retardation and it becomes possible to study the evolution of several mutually interacting vortex filaments with equations that are local in time. For instance, the motion of N vortex filaments is governed by the extrema of

$$\sum_{\alpha=1}^{N} \frac{F_\alpha}{3} \int dt\, d\sigma\, \mathbf{X}_\alpha \cdot \left(\frac{\partial \mathbf{X}_\alpha}{\partial t} \wedge \frac{\partial \mathbf{X}_\alpha}{\partial \sigma} \right)$$
$$+ \sum_{\alpha<\beta} \frac{f_\alpha f_\beta}{4\pi} \int dt\, d\sigma\, d\sigma' \left[\frac{1}{|\mathbf{X}_\alpha(\sigma, t) - \mathbf{X}_\beta(\sigma', t)|} \left(\frac{\partial \mathbf{X}_\alpha}{\partial \sigma} \cdot \frac{\partial \mathbf{X}_\beta}{\partial \sigma'} \right) \right]$$

In particular, two-dimensional point-vortex dynamics [12] is recovered with $\mathbf{X}_\alpha(\sigma, t) = (X^1_\alpha(t), X^2_\alpha(t), \sigma)$.

For a slightly compressible fluid, $\rho = \rho_0 + \rho_1(x)$ with $\rho_1 \ll \rho_0$, and if p is the pressure, $\nabla p = c^2 \nabla \rho_1$. The action (7) becomes

$$S \approx \rho_0 \int dt\, d^3x \left[\frac{\partial \phi}{\partial t} + \frac{1}{2} (\nabla \phi)^2 - \frac{1}{2} c^2 \left(\frac{\partial \phi}{\partial t} \right)^2 \right]$$

This time there is (sound) radiation and the evolution of a test vortex in a prescribed flow Φ is given by a differential equation only in a local approximation, when such radiation is neglected. The equation is the following:

$$\frac{\partial}{\partial t}\left(\frac{\partial L}{\partial \dot{\mathbf{X}}} \right) + \frac{\partial}{\partial \sigma} \left(\frac{L}{\partial \mathbf{X}'} \right) = (\nabla \Phi - \dot{\mathbf{X}}) \wedge \mathbf{X}' + \frac{1}{c^2} \frac{\partial \Phi}{\partial t} \dot{\mathbf{X}} \wedge \mathbf{X}'$$

with

$$L = f \ln\left(\frac{\delta}{\varepsilon}\right)\left[\left(1 - \frac{\dot{\mathbf{X}}^2}{c^2}\right)(\mathbf{X}')^2 + \frac{1}{c^2}(\dot{\mathbf{X}} \cdot \mathbf{X}')\right]^{1/2}$$

where an overdot denotes $\partial/\partial t$ and a prime denotes $\partial/\partial\sigma$. This equation is valid in the limit $\delta/\varepsilon \to \infty$, a very thin vortex whose radius of curvature is nowhere smaller than the relevant acoustic wavelengths. It provides the correction due to compressibility to the incompressible statement that a vortex is convected by a flow.

The ideas of this section can be used to compute the effect of finite source size on experimental results of vortex sound such as those of Kambe and Minota [13], where the usual formulation [14] might not be sufficiently accurate [15].

4. DISLOCATION LOOPS IN AN ELASTIC SOLID [16]

This time we have not a single scalar field but a (Euclidean) vector field $u^i(x, t)$, the displacement from their equilibrium position of points in a linearly elastic solid. The wave equation for this elastodynamic field is

$$\rho \frac{\partial^2 u^i}{\partial t^2} - c_{ijkl} \frac{\partial^2 u^k}{\partial x^j \partial x^l} = 0 \tag{8}$$

where ρ is the (uniform) density and c_{ijkl} is a tensor involving the elastic constants of the medium. If the latter is homogeneous and isotropic, the only case we shall consider here, $c_{ijkl} = \lambda\delta_{ij}\delta_{kl} + \mu(\delta_{ik}\delta_{jl} + \delta_{il}\delta_{jk})$, where λ and μ are Lamé coefficients, and there are two characteristic velocities corresponding to longitudinal and transverse waves. As in previous sections, it is possible to look for solutions of the wave equation that are not single valued but have a discontinuity, this time called the Burgers vector b^i, when crossing a surface $S(t)$ with boundary $B(t)$. The boundary is now called a dislocation loop and it is possible to ask, and answer, the same two questions we have asked for the relativistic string and the vortex filament: What is the radiation generated by a dislocation undergoing prescribed motion? What is the response of a dislocation to outside elastic radiation? The second one has been around for a long time* and an answer within the context of linear elasticity has become possible only now with the formalism that follows. The answer to the first one was found by Mura [18]: particle velocity $\partial u_i/\partial t$ and strain $\partial u_i/\partial x^k$ can be written as a convolution of the Green's function for equation (8) and source functions localized on the dislocation loop.

* See, for example, Ref. 17.

As in Sections 2 and 3, one notes that there is a variational principle from which the equations of elastodynamics follow: extremize

$$S = \frac{1}{2}\int dt\, d^3x \left[\rho\left(\frac{\partial u}{\partial t}\right)^2 - \frac{\partial u^i}{\partial x^k}\sigma_{ik}\right]$$

where $\sigma_{ik} = c_{ikjm}\,\partial u^j/\partial x^m$ is the elastic strain. The velocities and strain generated by a dislocation loop being singular on the loop, one considers the action integrated on a volume that excludes a thin tube around it, thereby becoming a functional of its world history whose extrema will determine the evolution of this stringlike dislocation. The main difficulty with carrying out this program is not one of principle but algebraic, due to the complicated nature of the relevant Green's function:

$$4\pi\rho G_{km}(\mathbf{x}, t) = \frac{\delta_{mk}}{\beta^2 r}\,\delta\left(t - \frac{|x|}{\beta}\right) + \frac{x_k x_m}{|x|^3}\left[\frac{1}{\alpha^2}\,\delta\left(t - \frac{|x|}{\alpha}\right) - \frac{1}{\beta^2}\,\delta\left(t - \frac{|x|}{\beta}\right)\right]$$
$$+\, t\left[\theta\left(t - \frac{|x|}{\alpha}\right) - \theta\left(t - \frac{|x|}{\beta}\right)\right]\left(\frac{3x_m x_k}{|x|^5} - \frac{\delta_{mk}}{|x|^3}\right)$$

Here θ is the step function and α and β are the longitudinal and transverse wave velocities. Contrary to the Green's function $G(\mathbf{x}, t) = (1/|x|)\delta(t - |x|/c)$ that has support only on the light cone, the Green's function G_{mk} has support on two "sound cones" as well as in the whole volume in between. Moreover, it is a matrix, reflecting the nonscalar nature of the elastic wave operator. Nevertheless, the full differential equation describing the response of a dislocation loop to an externally applied stress can be worked out. It is rather unwieldy and a detailed discussion is being reported elsewhere.* A particularly simple case is that of a screw dislocation; this is a two-dimensional problem in which only one component of displacement u^i is nonvanishing, the dislocation being now a point $\mathbf{X}(t)$. Its equation of motion is ($a = 1, 2$; $\varepsilon_{12} = -\varepsilon_{21} = 1$, $\varepsilon_{11} = \varepsilon_{22} = 0$)

$$M\frac{d}{dt}(\gamma\dot{X}_a) = \mu b\varepsilon_{ac}\frac{\partial U}{\partial x^c} + \rho b\varepsilon_{ac}\dot{X}_c\frac{\partial U}{\partial t} \tag{9}$$

with $M = (\rho b^2/4\pi)\ln(\delta/\varepsilon)$, and $U(x, t)$ the externally prescribed particle displacement. It is precisely the same equation obeyed by a uniformly charged straight line in classical electrodynamics. The left-hand side of (9) is the time derivative of a relativistic momentum, and the right-hand side is the analog of the Lorentz force.

ACKNOWLEDGMENTS. This work has been supported by grants from Departamento de Investigación y Bibliotecas de la Universidad de Chile (E-2489-8612) and Fondo Nacional de Ciencias (1147-1984).

* In the second work in Ref. 16.

REFERENCES

1. E. Siggia and A. Pumir, *Phys. Rev. Lett. 55*, 1749 (1985); R. H. Kraichnan, *Adv. Math. 16*, 305 (1975).
2. H. L. Swinney and J. P. Gollub (eds.), *Hydrodynamic Instabilities and the Transition to Turbulence*, 2nd Edition, Springer, New York, 1985. S. H. Davies and J. L. Lumley (eds.), *Frontiers in Fluid Mechanics*, Springer, New York, 1985.
3. M. E. Goldstein, *Aeroacoustics*, McGraw-Hill, New York, 1976.
4. R. Thomson, *Physics of Fracture*, to be published in the Solid State Physics series.
5. F. Lund, *Phys. Rev. D 33*, 3124 (1986).
6. R. Tabensky, *Phys. Rev. D 13*, 267 (1976).
7. Y. Nambu, *Phys. Rep. 23C*, 250 (1976).
8. E. Cremmer and J. Scherk, *Nucl. Phys. B72*, 117 (1974).
9. M. Kalb and P. Ramond, *Phys. Rev. D 9*, 2273 (1974).
10. F. Lund and T. Regge, *Phys. Rev. D 14*, 1524 (1976).
11. F. Lund and N. Zabusky, *Phys. Fluids* (to be published). F. Lund, *The Interaction of Sound and Vorticity*, in: *Instabilities and Non-Equilibrium Structures* (E. Tirapegui and D. Villaroel, eds.), Reidel, Dordrecht, 1987.
12. H. Aref, *Ann. Rev. Fluid Mech. 15*, 345 (1983).
13. T. Kambe and T. Minota, *Proc. R. Soc. London A386*, 277 (1983).
14. W. Möhring, *J. Fluid Mech. 85*, 685 (1978).
15. D. Risso and F. Lund, unpublished.
16. F. Lund, *Phys. Rev. Lett. 54*, 14 (1985); F. Lund, Response of a stringlike dislocation loop to an external stress, *J. Mater. Res.* (to be published).
17. F. R. N. Nabarro, *Theory of Crystal Dislocations*, Oxford University Press, Oxford, 1967; and J. P. Hirth and J. Lothe, *Theory of Dislocations*, McGraw-Hill, New York, 1982.
18. T. Mura, *Phil. Mag. 8*, 843 (1963).

Chapter 13

p-Form Monopoles

Claudio Teitelboim

I would like to report some results on an extension of electrodynamics, and in particular of the magnetic pole, in which the vector potential is replaced by a totally antisymmetric tensor with p indices, a p-form. The comments will be brief. Those interested may find more on the subject in Refs. 1-4.

It is natural to be interested in p-forms if one takes extended objects seriously. The reason is that just as a 1-form couples to a point particle through an action,

$$q \int A_\mu \, dz^\mu \tag{1}$$

a p-form couples to a $(p-1)$-dimensional object through

$$\frac{q}{p!} \int A_{\mu_1 \ldots \mu_p} \, dz^{\mu_1} \wedge \cdots \wedge dz^{\mu_p} \tag{2}$$

In equations (1) and (2), the integrals are taken over the history of the object. That history is denoted in parametrized form by its embedding in

CLAUDIO TEITELBOIM • Centro de Estudios Científicos de Santiago, Santiago 9, Chile; and Center for Relativity, The University of Texas at Austin, Austin, Texas 71782.

space-time $z(\xi)$. Thus in (1), the history is one dimensional, and there is only one ξ, whereas in (2) the history is p-dimensional and there are p ξ's.

Now, given the great success of Yang–Mills theory, it is tempting to try to develop nonlinear theories of p-forms by imitating the Yang–Mills construction, that is, by introducing a Lie algebra valued p-form

$$A_{\mu_1\ldots\mu_p} = A^a_{\mu_1\ldots\mu_p} t_a \tag{3}$$

However, this does not go through unless the gauge group is Abelian. This is why I started talking about electrodynamics instead of advertising an analog of chromodynamics.

When I say that the attempt does not go through, I have in mind a definite and no doubt not too imaginative context, namely, insisting that the geometrical meaning of a 1-form is that of a connection that describes parallel transport along a line. The extension of this idea to a p-form would be to attempt describing transport along a p-dimensional manifold.

But the idea of transport involves a succession of group actions that are performed one after the other as one proceeds. This requires a notion of order defined on the manifold along which the transport is performed. Such a notion exists between the points of a one-dimensional line, but it does not exist between submanifolds of dimension $p-1$ immersed in a common p manifold, for p greater than or equal to 2. Therefore, one must transport with a group for which the order of the factors is immaterial. If one demands, in addition, that the group be compact, the only possibility is $U(1)$.

For $U(1)$ the generator t_a in (3) may be replaced by the imaginary unit i. The p-form connection A describes then parallel transport of complex functional fields defined over the space of $(p-1)$-dimensional surfaces. If one denotes by $x_1(\xi)$ and $x_2(\xi)$ two parametrized $(p-1)$-dimensional surfaces, the field $\phi[x_2(\xi)]$ obtained by parallel transport from x_1 to x_2 along a p-manifold Γ_p whose boundaries are x_1 and x_2 is

$$\phi_\Gamma[x_2(\sigma)] = \exp\left(-iq\int_{\Gamma_p} A\right)\phi[x_1(\sigma)] \tag{4}$$

where q is a coupling constant with dimensions of a length raised to the $(\frac{1}{2})[n-2(p+1)]$th power. Here n is the space-time dimension.

I would like to discuss next what is the analog of both electric and magnetic charge in this kind of theory.

The source of the field is now a p-form current j, which is conserved

$$\partial_{\mu_1} j^{\mu_1\ldots\mu_p} = 0 \tag{5}$$

or equivalently

$$d\,{}^*j = 0 \tag{6}$$

One might feel that there will be many conserved charges because there are many free indices in (5), so that one would obtain one conserved charge $Q^{\mu_1 \cdots \mu_{p-1}}$ by integrating $j^{0\mu_1 \cdots \mu_{p-1}}$ over all space. However, this is incorrect. Indeed, the charges defined that way would be either zero or infinity. This may be seen without calculation by observing that the resulting charge would be a (totally antisymmetric) space-time tensor whose time component would vanish in all Lorentz frames. This can only happen if all the components vanish or if at least one component is infinite, in which case the charge is not defined. The underlying reason is that for an antisymmetric tensor $\partial_{\mu_1} j^{\mu_1 \cdots \mu_p} = 0$ implies $\partial_{m_1} j^{0 m_1 \ldots m_{p-1}} = 0$, where the m's are purely spatial.

There was of course no way to obtain many charges. There should be only one total charge, because we are dealing with the group $U(1)$, which has only one generator. That total charge is provided by starting from the form (6) of current conservation. Upon integration over an $(n - p + 1)$-dimensional manifold M_{n-p+1} with boundary $(\partial M)_{n-p}$, and using Stoke's theorem, (6) we obtain

$$\int_{(\partial M)_{n-p}} {}^*j = 0 \tag{7}$$

Now, one may consider ∂M to be formed by two spacelike $(n - p)$ surfaces Σ_{n-p} and a remote timelike tube. If the current j is localized in space, the integral over the remote tube vanishes and one finds from (7) that the charge

$$Q = \int_{\Sigma_{n-p}} {}^*j \tag{8}$$

is independent of the spacelike section Σ_{n-p}. This implies that Q is Lorentz invariant and conserved.

In the particular case $p = 1$, (8) gives the usual integration over an $(n - 1)$-surface of the component of j normal to that surface. [The normal comes in from the dual in (8).] One sees, however, that the higher the dimension of the extended object the lesser the dimension of the cut Σ_{n-p} that comes into the definition of the total charge.

One can also find an expression for Q in terms of the field strength F which is a generalization of Gauss's law in electromagnetism, which reads, in this case,

$$d\,{}^*F = (-1)^{(n-p-1)}\,{}^*j \tag{9}$$

where

$$F = dA \tag{10}$$

is the field strength. Therefore, if one denotes by $(\partial\Sigma)_{m-p-1}$ the $(n-p-1)$-dimensional boundary of Σ_{n-p}, one has the expression

$$Q = (-1)^{(n-p-1)} \int_{(\partial\Sigma)_{n-p-1}} {}^{*}F \tag{11}$$

for the total charge within $(\partial\Sigma)_{n-p-1}$.

For an extended object of dimension $p-1$ with a "p-dimensional world line" $z(\xi)$, the current takes the form

$$j^{\mu_1 \cdots \mu_p}(x) = q \int \delta^{(n)}(x - z)\, dz^{\mu_1}(\xi) \wedge \cdots \wedge dz^{\mu_p}(\xi) \tag{12}$$

In that case, contributions to (8) come only from the intersection of Σ_{n-p} with the p-dimensional history of the object. That intersection has dimension zero and consists, therefore, of a number of points with each point giving a contribution $\pm q$.

The preceding discussion of the total charge parallels closely that for electrodynamics. However, an important additional property appears for extended objects. It is this: the total charge of an object that is not infinitely extended in all its directions is equal to zero. Thus, the analog of a single point charge in electrodynamics is an infinitely extended object. On the other hand, a compact object is the analog of a particle–antiparticle pair.

This property follows also from the divergence-free character of the current. In fact, if one evaluates the divergence of (12) by differentiation under the integral sign, one finds

$$\partial_{\mu_1} j^{\mu_1 \cdots \mu_p}(x) = -q \int d[\delta^{(n)}(x - z(\xi))\, dz^{\mu_2} \wedge \cdots \wedge dz^{\mu_p}] \tag{13}$$

where the exterior derivative acts on z. Next, one rewrites the right-hand side of (13)—which is an integral over the history of the object—as an integral over the history's boundary. It follows that, since (13) must vanish for every x, the history must have no boundary. This is an alternative statement of the conservation of charge in this formalism.

Now, if the history has no boundary, it must be either compact or infinitely extended in each direction. If there is at least one spatial direction in which the history is compact, one may find a boundary $(\partial\Sigma)_{n-p-1}$ located at a finite distance that totally surrounds the object. Then the intersection of Σ_{n-p} with the history consists of an even number of points, all of them lying in the interior of $(\partial\Sigma)_{n-p-1}$ and with the $(+q)$ contributions exactly canceling the $(-q)$ ones.

This situation is similar to what happens with pair annihilation in electrodynamics of point charges. There, one has an electron first moving forward in time and then coming back as a positron. The electron describes a continuous history that crosses each spacelike plane either twice or never, so that the total charge is conserved and equal to zero.

Here, given the spatially extended nature of the object, the same phenomenon happens in spacelike directions. If the object is localized, for each time that the history crosses Σ_{n-p} with a $(+q)$ contribution, there is a "return" piece that gives $(-q)$. If the object is infinitely extended in all directions, some of the return portions may be thought of as having moved out to infinity, leaving an unbalanced net contribution to the total charge.

So far, what I have discussed is the straightforward extension of the usual point particle electrodynamics to its p-form analog. Now I would like to show how to incorporate the analog of the magnetic pole. First of all, it should be said that this would be the analog of the Dirac magnetic pole and not of the t'Hooft-Polyakov object. Because the group is $U(1)$ this monopole will be a singular object. As singular, so to speak, as the electric pole. Now, in the case of strings maybe one is not too unhappy saying that it would be as singular as a string, because the string seems to be a nice object. So it seems that one should not apologize for the fact that one would have an analog of the Dirac pole and not of the t'Hooft-Polyakov monopole.

The idea is very simple and I would like to present it, because it is faster, with the help of the so-called Dirac string, which will of course have nothing to do with the superstring. It can also be done in a more sophisticated way with overlapping patches of different gauges [3], but I will present it here with the help of the Dirac string.

One simply extends in a straightforward way what Dirac did in 1948 [5]. I will call an electric pole the original object that has the current j given by (12), so the electric pole itself is $p-1$ dimensional. Then the magnetic pole history will have dimension $n-p-2$. Why? Because the field strength will have dimension $p+1$ and we want some potential that would be appropriate for the dual, so the potential should be an $n-p-2$ form and the dimension of the magnetic pole itself will be $n-p-3$.

Now, the Dirac string must be such that it is also an extended object whose boundary is the magnetic pole, so the Dirac string will have dimension $n-p-2$ and its history will have dimension $n-p-1$. Thus, the Dirac string is a kind of flag attached to the magnetic pole. A quick check shows that if $n=4$ and $p=1$ the Dirac string turns out to be of dimension 1, which is right and is why it was called "string" in the first place.

The next step is to introduce the current associated with the Dirac string. If one denotes by $\bar{y}(\bar{\xi})$ the parametric equations of the Dirac string history (there will be $n-p-1$ $\bar{\xi}$'s), that current will read

$$G^{\mu_1\cdots\mu_{n-p-1}}(z) = \bar{q}\int \delta^{(n)}(x-\bar{y})\, d\bar{y}^{\mu_1}\wedge\cdots\wedge d\bar{y}^{\mu_{n-p-1}} \tag{14}$$

Here $\bar{q}$ is a coupling constant, which will be identified below with the strength of the magnetic pole. This equation has the same form as (9) for the electric current, but it is not conserved because the surface $\bar{y}(\bar{\xi})$ has a boundary. One finds because of this fact

$$\partial_\mu G^{\mu\mu_1\cdots\mu_{n-p-2}} = -\bar{j}^{\mu_1\cdots\mu_{n-p-2}} \tag{15}$$

where

$$\bar{j}^{\mu_1\cdots\mu_{n-p-2}}(x) = \bar{q}\int \delta^{(n)}(x-\bar{z})\, d\bar{z}^{\mu_1}\wedge\cdots\wedge d\bar{z}^{\mu_{n-p-2}} \tag{16}$$

is the magnetic current. Here $\bar{z}(\bar{\xi})$ (with $n-p-2$ $\bar{\xi}$'s) are the parametric equations of the magnetic pole history which is the boundary of the surface $\bar{y}(\bar{\xi})$, representing the history of the Dirac string.

In the presence of the magnetic pole one modifies the definition (10) of the field strength to read

$$F = dA + (-1)^{(p+1)(n+1)}\,{}^*G \tag{17}$$

which implies

$$\partial_\mu\,{}^*F^{\mu\mu_1\cdots\mu_{n-p-2}} = \bar{j}^{\mu_1\cdots\mu_{n-p-2}} \tag{18}$$

and one sees that the magnetic current has been brought in as a source for ${}^*F$.

The complete set of equations of motion for the electric poles, the magnetic poles, the p-form field and the Dirac strings, may be derived from an action principle which again is the straightforward extension of the one given by Dirac [5]. One finds that the Dirac string has no equation of motion and can be located anywhere but is subject to the so-called "Dirac veto," which says that the string cannot intersect an electric pole.

Incidentally, this "Dirac veto" eliminates an interesting possibility that would otherwise have been available for $p > 1$. It is the idea of a "topologically stable dyon," which would consist of an electric and a magnetic pole linked with each other. Such a configuration would violate the "Dirac veto."

If one demands that the Dirac string should remain a pure gauge object in quantum mechanics, one finds again here that the product of the electric and magnetic strengths must be quantized:

$$q\bar{q} = 2\pi\hbar\cdot m \qquad (m \text{ an integer}) \tag{19}$$

Note that the product of the electric and magnetic couplings is always a pure number, although q and $\bar{q}$ separately have that property only when $n = 2(p+1)$.

ACKNOWLEDGMENTS. I would like to thank Marc Henneaux for many discussions. This work was supported in part by a grant of The Tinker Foundation to the Centro de Estudios Cientificos de Santiago and the National Science Foundation under grant No. PHY-8216715 to The University of Texas at Austin.

REFERENCES

1. C. Teitelboim, Gauge invariance for extended objects, *Phys. Lett. 167B*, 63 (1986).
2. C. Teitelboim, *Monopoles of Higher Rank, Phys. Lett. 167B*, 69 (1986).
3. M. Henneaux and C. Teitelboim, *p*-Form electrodynamics, in honor of Professor John Archibald Wheeler on his 75th birthday, *Found. Phys. 16*, 593 (1986).
4. M. Henneaux and C. Teitelboim, Quantization of topological mass in the presence of a magnetic pole, *Phys. Rev. Lett. 56*, 689 (1986).
5. P. A. M. Dirac, The theory of magnetic poles, *Phys. Rev. 74*, 817-830 (1948).

Chapter 14

The Gravitational Path Integral and Critical Dimensions of Linear and Nonlinear Locally Supersymmetric σ Models

Peter van Nieuwenhuizen

1. INTRODUCTION

It is well known that Slavnov–Taylor identities can be derived from path integrals by making a change of integration variable which corresponds to a symmetry transformation and using the fact that the classical action is invariant under this symmetry. When the measure is invariant under the symmetry transformation one obtains only contributions from the action and external sources. For rigid symmetries, one first replaces the parameter by a local parameter and in this way the Noether current enters in the Ward identities. For local symmetries with external gauge fields one can make a local classical gauge transformation, but if the gauge fields are dynamical, one must instead consider the rigid BRST symmetry of the quantum action.

When the measure is not invariant under the symmetry considered, it yields according to Fujikawa [1] extra terms to the Slavnov–Taylor identities, the so-called anomalies. The anomaly is then the product of all Jacobians at all space-time points. One may regulate this ill-defined infinite product in various ways, so as to make it well defined. Three regularizations we will consider are the following:

PETER VAN NIEUWENHUIZEN • Institute for Theoretical Physics, State University of New York at Stony Brook, Stony Brook, New York, 11794.

1. One expands the Jacobian into a complete set of eigenfunctions, and truncates the infinite sum. This regularization we shall call mode cutoff.
2. The exponential regularization consists of adding a regulator $\exp(R/M^2)$ with eigenvalues $R = -\lambda_m^2$ and taking at the end of the calculation the limit $M^2 \to \infty$. This regularization is easier to compute because one can use a complete set of plane waves to evaluate traces, in which case the integration over the complete set often becomes a Gaussian integral over k.
3. The zeta-function regularization consists of replacing such Green's functions as $\sum \phi_m(x)\phi_m(y)$ by $\sum \phi_m(x)\phi_m(y)\lambda_m^s$ and letting s go to zero at the end of the computation. In the Gaussian regularization one uses instead $\exp(-\lambda_m^2/M^2)$. By rewriting λ_m^s as a Mellin integral over t divided by $\Gamma(s)$ one finds for s to zero that only $t = 0$ contributes and yields the same result as Gaussian regularization, except that no s^{-k} divergences remain.

In this way one obtains the correct axial anomalies [2], trace anomalies [3], and critical dimensions of linear σ models (spinning strings) [4-6], and of nonlinear σ models [27, 33] and also the anomalies for higher derivative field theories [7], and in superspace [40]. One may also *define* the anomaly as the product of the Jacobians at all points if the original action is not invariant, for example, in the case of trace anomalies for an ordinary (nonimproved) scalar. In such cases, these anomalies due to the measure are different for fields that are classically the same [8] (for example, a scalar and a two-index antisymmetric tensor in four dimensions), while the "total anomaly" (the sum of the contributions from the measure and the noninvariant classical action) is the same [9]. If the contribution from the measure has a physical meaning of its own (something that is not known), one has "quantum inequivalence of classically equivalent theories" [8]; otherwise one has "quantum equivalence etc." [9].

Before going on we must specify what we mean by measure. We begin by noting that there is freedom in the choice of integration variables of a path integral. For example, for a scalar one may use the original field $S(x)$ and write the measure as $DS(x)$, but we shall argue that a better choice is instead $\tilde{S}(x) \equiv e(x)^{1/2}S(x)$ and use as measure $D\tilde{S}(x)$. Here e_μ^m denotes the vielbein field and e its determinant. Thus measure means first of all: choice of independent integration variable. In addition to choosing the independent integration variables, one must also choose a regulator. It is only the *pair* of integration variable plus regulator that constitutes "the measure." The actions $I(S, e)$ and $I(\tilde{S}, e)$ depend in different ways on e_μ^m, and hence if the regulator is obtained from the action by some prescription then different integration variables will in general lead to different Jacobians and to different anomalies. Although such variables as $\tilde{S}(x)$ can always be written as the product of $S(x)$ and $e^{1/2}$, we shall treat the integration variables as the basic independent variables and write the action and other quantities in terms of the independent variables.

Note how useful this notion of freedom in the choice of integration variable is. Suppose one considers a vector field in four dimensions coupled to an external gravitational field. If one uses $A_\mu(x)$ as independent variable, there is no local Weyl anomaly because the classical action is Weyl invariant under $\delta A_\mu = 0$, $\delta g_{\mu\nu} = \Lambda(x) g_{\mu\nu}(x)$, so that also the measure DA_μ is Weyl invariant. There is then, however, a local coordinate anomaly as we shall discuss. On the other hand, taking as basic integration variable $\tilde{A}_\mu = A_\mu e^\alpha$ with $\alpha = \frac{1}{2}(d-2)/d$, there *is* a Weyl anomaly but no coordinate anomaly. The freedom in the measure thus has physical implications, namely, different anomalies, and one can use this freedom to eliminate some (but not all) anomalies.

It would be desirable to decide upon the correct measure by requiring that divergences in loop graphs cancel, in the same way as in nonlinear σ models the $\delta^4(0)$ measure cancels quartic divergences. That would give the measure a direct meaning in terms of Feynman diagrams. This program we intend to execute in the future, but for the time being another route has been followed.

In order to fix the measure one requires that it is invariant under some preferred symmetries; these are then the symmetries without anomalies. Having fixed the measure in this way, one may then proceed to calculate the anomalies for other symmetries. Different choices of the measure (i.e., different choices of preferred symmetries) will in general lead to different anomalies for the nonpreferred symmetries. The choice of preferred symmetries is a matter of physical prejudice and is analogous to the requirement of vector conservation in Yang-Mills theories which then leads to axial vector anomalies.

For systems coupled to an external gravitational field, one chooses a measure that is invariant under reparametrizations [10]. This is analogous to the case of external Yang-Mills fields coupled to fermions with a measure that is invariant under vector transformations. As is well known [11], one can always remove a reparametrization anomaly by shifting it to a local Lorentz anomaly, just as one can shift the vector anomaly in Yang-Mills theories to an axial vector anomaly.

When the gravitational field itself is quantized, the natural symmetry to impose is BRST coordinate symmetry, which is the residual rigid symmetry of the quantum action that remains after the local classical gauge invariance of reparametrizations has been fixed and the coordinate ghosts have been introduced. Since the coordinate ghosts and antighosts serve the purpose of removing the unphysical degrees of freedom of the graviton at the quantum level, their measure should be treated simultaneously with that of the graviton.

In theories with fermions, the gravitational field is usually described by the vielbein field $e_\mu^m(x)$. In addition to local coordinate invariance, one has then at the classical level also another gauge invariance, namely, local Lorentz invariance. One should also add for this symmetry a gauge-fixing term and an ensuing ghost action, and one should again treat the measures of the Lorentz ghosts and antighosts simultaneously with that of the vielbein and coordinate ghosts and antighosts. The natural gauge-fixing term for local Lorentz invariance is the square of the antisymmetric part of the vielbein [12]

$$\mathscr{L}(\text{Lorentz fixing}) = \alpha(e_\mu^m - \delta_\nu^m \delta_\mu^n e_\nu^n)^2 \tag{1}$$

with α a constant. The Lorentz ghosts and antighosts are then not propagating.

Instead of starting with local Lorentz invariance at the classical level and then treating it on a par with other local symmetries by means of the covariant quantization techniques, one could begin by choosing a suitable "unitary" gauge in the classical theory. The most natural such gauge is the "symmetric gauge" in which the vielbein is symmetric. Other local classical symmetries, such as reparametrizations, will in general break this symmetry and one has to add field-dependent compensating local Lorentz transformations which bring one back in the symmetric gauge. It has been shown in Ref. 13 that quantization and classical gauge fixing are commuting operations: Integrating out the Lorentz ghosts and antighosts, the Lorentz ghosts cannot just be dropped, but they become equal to a certain combination of other fields and lead in this way to extra interactions. These same extra interactions are obtained if one applies the standard covariant quantization methods to the "corrected" classical symmetries (corrected in the sense that they maintain the symmetric gauge). In what follows we shall not work in the symmetric gauge, and treat local Lorentz symmetry on the same footing as a local Yang-Mills symmetry.

Having chosen for each field that measure (i.e., independent integration variable *plus* regulator) that is free from coordinate anomalies, one can then compute the local Lorentz anomaly unambiguously. One finds (not surprisingly) that only theories with chiral fermions can have local Lorentz anomalies.

For supersymmetric theories the requirement of absence of coordinate anomalies of each field separately of a given multiplet does not uniquely fix the measure. One can still take *linear combinations* of matter fields multiplied by vielbeins *and gravitinos.* Imposing the requirement that also local supersymmetry anomalies are absent then fixes the measure completely. For a discussion of these aspects see Ref. 39.

We now give a brief discussion of the Becchi–Rouet–Stora–Tyutin formalism. The BRST symmetry for gauge theories is a quantum extension of classical gauge symmetry. The classical gauge action plus gauge-fixing terms plus Faddeev–Popov ghost action has a rigid symmetry with a constant anticommuting parameter Λ, which is an extension of the classical local gauge invariance of the classical gauge action.

Let the classical gauge fields be generically denoted by Φ^i, and let the classical gauge transformations be given by

$$\delta\phi^i = R(\phi)^i_\alpha \xi^\alpha \tag{2}$$

The indices i and α run over internal as well as space-time variables. For example, in Yang-Mills theory $\delta W^a_\mu(x) = (D_\mu \Lambda)^a$, where $\{a, \mu, x\}$ constitute i, and $\{a, x\}$ correspond to α. To each local parameter ξ^α one associates a ghost field C^α with opposite statistics. Then the BRST transformation rule of the classical gauge fields is

$$\delta\phi^i = R(\phi)^i_\alpha C^\alpha \Lambda \tag{3}$$

Taking ξ^α and C^α real, it follows that Λ is imaginary since $(C^\alpha \Lambda)^* = \Lambda^* C^\alpha$.

We shall assume that the classical gauge algebra closes. That means that the commutator of two local gauge transformations is a sum of local gauge transformations

$$\begin{aligned}[\delta(\eta), \delta(\xi)]\phi^i &= R^i_{\alpha,j} R^j_\beta \eta^\beta \xi^\alpha - \xi \leftrightarrow \eta \\ &= R^i_\gamma f^\gamma_{\alpha\beta} \eta^\beta \xi^\alpha \end{aligned} \tag{4}$$

The symbol $R^i_{\alpha,j}$ denotes the right derivative of R^i_α with respect to ϕ^j. This formula *defines* the structure constants, which may be (and are in the case of supergravity theories) field dependent. The BRST transformation laws of the classical gauge fields ϕ^i are then nilpotent if the ghost fields C^α transform under BRST transformations as follows:

$$\delta C^\alpha = -\tfrac{1}{2} f^\alpha_{\beta\gamma} C^\gamma \Lambda C^\beta \tag{5}$$

In this result and other results below we shall use equations that are valid for bosonic and fermionic local symmetries simultaneously [14].

The nilpotency of BRST transformations on gauge fields ϕ^i follows from the closure of the classical gauge algebra. Nilpotency on the ghosts follows from the Jacobi identities for three local gauge transformations [14]. When the structure "constants" are field dependent, the Jacobi identities are modified and contain terms with derivatives of $f^\alpha_{\beta\gamma}$ [16] but the transformation rules in (3) and (5) are general and hold in these cases as well. This is the geometrical part of BRST transformations.

To complete the BRST transformations, one adds auxiliary fields that are BRST inert:

$$\delta d^\alpha = 0 \tag{6}$$

The antighosts transform per definition into the auxiliary fields

$$\delta C^{*\alpha} = -d^\alpha \Lambda \tag{7}$$

Obviously, the complete BRST transformation rules are still nilpotent as long as the fields commute or anticommute (i.e., neglecting normal ordering effects).

The quantum action is given by

$$\begin{aligned}\mathscr{L}(\text{quantum}) = \mathscr{L}(\text{class}, \phi^i) - (-)^\alpha d^\alpha \gamma_{\alpha\beta}(F^\beta + \tfrac{1}{2}d^\beta) \\ + C^{*\alpha}\gamma_{\alpha\beta}F^\beta_{,\gamma}C^\gamma\end{aligned} \tag{8}$$

if $\gamma_{\alpha\beta}$ is field independent. (There is no sign in the last term since C^α and $C^{*\alpha}$ always commute with F_α.) The symbol $F^\beta_{,\gamma}\zeta^\gamma$ means the gauge variation of F^β with parameter ζ^γ. Sometimes one uses field-dependent $\gamma_{\alpha\beta}$; in that case [15, 16]

$$\mathscr{L}(\text{quantum}) = \mathscr{L}(\text{class}, \phi^i) + (\delta/\delta\Lambda)[C^{*\alpha}\gamma_{\alpha\beta}(F^\beta + \tfrac{1}{2}d^\beta)] \tag{9}$$

where $\delta/\delta\Lambda$ indicates first making a BRST variation and then removing Λ from the right.

One can even construct a theory of *local* BRST invariance in superspace [16] but that seems not of use for string theories.

Let us now discuss how the measure is determined by requiring BRST coordinate invariance.

2. THE CORRECT MEASURE

For gravitational theories, the correct integration variables are not the original variables but rather products of the original variables with powers of the determinant of the vielbein e_μ^m. For example, for a scalar field, in any dimension, the correct integration variable is

$$\tilde{S}(x) = e(x)^{1/2}S(x) \tag{10}$$

The proof that this measure is indeed invariant under classical coordinate transformations (when the gravitational field is external), or quantum (BRST) coordinate transformations when one also integrates over the gravitational field is rather simple and will be given below. As regulator one may use in this case the Klein-Gordon operator on the basis with $\tilde{S}$ as we shall discuss,

$$R = e^{-1/2}\partial_\mu e g^{\mu\nu}\partial_\nu e^{-1/2} \tag{11}$$

or simply mode cutoff. Both regulators can be used for $\tilde{S}$ and lead to absence of coordinate anomalies.

For other classical fields such as spinors or vectors, one may first convert them to tensors with flat indices which are scalars under coordinate transformations, after which the same considerations as for scalars are applicable. For example, for vectors A^μ, the corresponding tensors with flat indices are $A^m \equiv A^\mu e_\mu^m$ and thus the correct integration variable is given by

$$\prod_m D\tilde{A}^m = \prod_m D(e^{1/2} e_\mu^m A^\mu) \tag{12}$$

In applications one often works with tensors with curved indices, and one may determine their measure as follows:

$$\prod_m D(\tilde{A}^m) = e^{d/2} e \prod_\mu (DA^\mu) = \prod_\mu D(e^{(d+2)/2d} A^\mu) = \prod_\mu (D\tilde{A}^\mu) \tag{13}$$

However, for rigorous proofs one should always go back to the variable $\tilde{A}^m$. We shall not discuss matter fields any further.

When one comes to the correct measure of the vielbein and the coordinate ghost field, there arises a problem. Since ghosts serve the purpose of eliminating the nonphysical degrees of freedom of the gauge field, their measure should be treated simultaneously with that of the gauge field. However, under BRST transformation, coordinate ghosts do not transform as vectors; rather they transform as follows:

$$\delta C^\mu = C^\nu \Lambda \partial_\nu C^\mu \tag{14}$$

[An easy way to check this result is to use nilpotency of coordinate BRST transformations on an ordinary scalar field $S(x)$.] If one evaluates straightforwardly the BRST Jacobian of the vielbein and ghost fields, one finds, in fact, terms with derivatives of the vielbein that are absent in the BRST Jacobian of matter fields and it seems that the Jacobian is not a total derivative [17]. The reason these extra terms arise can be traced back to the fact that ghosts are not good tensors, as we already said. We shall perform this calculation explicitly.

One might think that by clever rearrangement of all these extra terms with derivatives of the metric one might still be able to show that the BRST coordinate Jacobian of the vielbein-ghost system is a total derivative. In that case, one would have shown that there are no coordinate anomalies. Inspection of the ∂e terms (i.e., with vielbein derivatives) shows that this is not the case: for no choice of independent integration variable of the vielbein and coordinate ghost does their combined Jacobian become a total derivative. This might seem the end of the road; however, there is a solution, namely, the ∂e terms are ambiguous [18]. We shall show that one can perform the computation in (at least) two ways and get different answers. These related features, spurious ∂e terms and ambiguity, are present before regularization, and regularization may make matters worse. (By this we mean that certain naive operations that were used in the absence of regulators, may actually not hold in the presence of regulators.)

Our strategy will thus be to first *define* the unregularized Jacobian, then regularize it, and only then prove the absence of BRST coordinate anomalies. This definition of the Jacobian of the vielbein–coordinate ghost system should then also be used in further applications, whenever one computes the contribution of the measure to a Slavnov–Taylor identity.

The key to our definition of Jacobian and subsequent proof of absence of local coordinate anomalies is the observation that one can define the Jacobian as a product of three Jacobians, each of which has a clear meaning [18]. This proposed rewriting leads naturally to a proof of absence of coordinate anomalies, but it should be stressed that the very fact of assuming that this factorization of the Jacobian is valid in the presence of regularization already is part of our fixing of the ambiguity.

We rewrite the original Jacobian as a product, such that this product splits into a vielbein and a coordinate ghost part for each of which the Jacobian is a total derivative. The factorization of the Jacobian in the vielbein–coordinate ghost system is given by [18]

$$\begin{aligned} J &= \operatorname{sdet} \frac{\partial\{\tilde{e}_\mu^m(x,\Lambda), \tilde{C}^m(x,\Lambda)\}}{\partial\{\tilde{e}_\nu^n(y), \tilde{C}^n(y)\}} \\ &= \operatorname{sdet} \frac{\partial\{\tilde{e}_\mu^m(x,\Lambda), \tilde{C}^m(x,\Lambda)\}}{\partial\{\tilde{e}_\nu^n(y,\Lambda), C^\nu(y)\}} \\ &\quad\times \operatorname{sdet} \frac{\partial\{\tilde{e}_\nu^n(y,\Lambda), C^\nu(y)\}}{\partial\{\tilde{e}_\rho^r(z), C^\rho(z)\}} \\ &\quad\times \operatorname{sdet} \frac{\partial\{\tilde{e}_\rho^r(z), C^\rho(z)\}}{\partial\{\tilde{e}_\sigma^s(t), \tilde{C}^s(t)\}} \end{aligned} \tag{15}$$

Our notation is as follows:

$$\tilde{e}_\mu^m = e^k e_\mu^m, \qquad \tilde{C}^m = e^l e_\mu^m C^\mu \tag{16}$$

with $e = \det e_\mu^m$ and k and l constants to be determined such that J becomes unity. Further $\tilde{e}_\mu^m(x,\Lambda)$ equals the variation of $\tilde{e}_\mu^m(x)$ under a symmetry transformation, in this case a BRST coordinate transformation with Λ the constant anticommuting imaginary BRST parameter. To see the meaning of this factorization, observe that the one-but-last Jacobian is simply the Jacobian of $\tilde{e}_\mu^m(x)$ keeping the coordinate ghost field $C^\mu(x)$ with curved index fixed. This is the natural way of writing a BRST variation of the vielbein

$$\delta e_\mu^m = C^\nu \Lambda \partial_\nu e_\mu^m + (\partial_\mu C^\nu \Lambda) e_\nu^m \tag{17}$$

Naively, therefore, the treatment of *this* Jacobian is expected to be similar to that of a classical matter field. However, when one tries to give a better treatment involving a complete set of functions for the vielbein, one finds that the curved index of the vielbein gives various problems [41]. Obviously one cannot make this index flat by multiplication with an inverse vielbein (at least, not if one is not working in a background field formalism), and this difference with classical fields requires a drastic new definition of the Jacobian of the vielbein [18]. Essentially, one defines the Jacobian as some averaged Jacobian, averaged over $\tilde{e}_\mu^m$ and $\tilde{e}_m^\mu$ such that the effects of the curved index disappear. For pedogical reasons we will ignore the problems of the curved index due to regularization, and present a naive derivation of the correct measure for the vielbein. The result is

$$\tilde{e}_\mu^m(x) = e(x)^{(d-2)/2d} e_\mu^m(x) \tag{18}$$

and agrees with the more rigorous derivation in [18] and with the original proposal of Fujikawa in [10] and with the simple algorithm in (13).

The two remaining Jacobians in (15) also have a clear meaning. Note that they are evaluated at different points, at $\tilde{e}_\mu^m(x, \Lambda)$ and at $\tilde{e}_\mu^m(x)$. This "point-splitting in BRST space" amounts to rearranging the term in the original ill-defined Jacobian such that they can be divided into terms dealing with the measure of the vielbein field (whose Jacobian we just discussed) and terms that deal with the measure of the ghost field, which we shall now discuss. The latter kind of terms are given by

$$\operatorname{sdet}\left[\frac{\partial \tilde{C}^m(x,\Lambda)}{\partial C^\mu(y)}\right]_{\tilde{e}_\mu^m(x,\Lambda)} \operatorname{sdet}^a\left[\frac{\partial C^\mu(y)}{\partial \tilde{C}^n(z)}\right]_{\tilde{e}_\nu^n(y)} \tag{19}$$

One can write this product of determinants as a determinant of products. Since we have then a matrix in the space of ghost fields with *flat* indices $\tilde{C}^m(x)$, we can expand $\tilde{C}^m(x)$ into a complete set of functions analogously to the expansion of a scalar, and the previously mentioned terms with derivatives of the vielbein disappear. Moreover, this mode expansion allows us to regulate the Jacobian by mode cutoff, i.e., by truncating the finite sum over all modes, and still maintaining a unit Jacobian.

At this point one may better appreciate what the rewriting of the original Jacobian has achieved. We have diagonalized the problem into a vielbein $\tilde{e}_\mu^m(x)$ sector and a ghost $\tilde{C}^m$ sector, each of which can still be regularized by a different regulator. Before the rewriting no such diagonalization was present, and each determinant would have required its own $\tilde{e}$- *and* $\tilde{C}$-dependent regulator.

We now provide the announced details. First we consider a scalar field $S(x)$. Under BRST coordinate transformations one has

$$e^q(x, \Lambda)S(x, \Lambda) = e^q S + C^\rho \Lambda \partial_\rho (e^q S) + q(\partial_\rho C^\rho \Lambda) e^q S \tag{20}$$

where $C^\mu(x, \Lambda)$ and $e^m(x, \Lambda)$ were given in (14) and (17). The Jacobian for $\tilde{S}(x)$ reads

$$J = \det[\partial \tilde{S}(x, \Lambda)/\partial \tilde{S}(y)] = \det\{\delta(x-y) \\ -\Lambda[C^\rho(x)\partial^x_\rho \delta(x-y) + q(\partial^x_\rho C^\rho(x))\delta(x-y)]\} \tag{21}$$

(It obviously does not matter in this Jacobian whether we keep C and e or $\tilde{C}$ and $\tilde{e}$ fixed.) Since Λ is nilpotent, we only get terms linear in Λ in the expansion:

$$J = 1 - \Lambda \int dx\, dy\, \delta(x-y)[C^\rho(x)\partial^x_\rho \delta(x-y) \text{ etc.}] \tag{22}$$

Using that $[\partial^x_\rho \delta(x-y)]|_{x=y}$ is equal to $\frac{1}{2}\partial^x_\rho[\delta(x-y)|_{x=y}]$ we obtain a total divergence [10] for $q = \frac{1}{2}$

$$J = 1 - \Lambda \int dx\, \partial^x_\rho\{[C^\rho(x)\delta(x-y)]_{x=y}\} \tag{23}$$

It may be clearer to represent the delta function as an infinite sum over a complete set of $\tilde{\phi}_m$ belonging to the self-adjoint operator R in (11) and satisfying $\tilde{\phi}_m = e^{1/2}\phi_m$, with $\tilde{\Box}\tilde{\phi}_m = -\lambda^2_m \tilde{\phi}_m$ and $\int \tilde{\phi}_m \tilde{\phi}_n\, dx = \delta_{mn}$. In that case we can rewrite the delta function as

$$\delta(x-y) = \sum_m \tilde{\phi}_m(x)\tilde{\phi}_m(y) \tag{24}$$

and we may use that for fixed m

$$(C^\rho(x)\partial^x_\rho \tilde{\phi}_m(x)\tilde{\phi}_m(y))_{x=y} = C^\rho(x)\partial^x_\rho(\tfrac{1}{2}\tilde{\phi}_m(x)\tilde{\phi}_m(x)) \tag{25}$$

By proceeding as before one again obtains a total derivative, but now even if the summation over m is truncated. Hence, even in the presence of mode cutoff regularization, the Jacobian of a scalar equals unity. Also for exponential regularization, the Jacobian remains unity because the *constant* factors $\exp(-\lambda^2_m/M^2)$ can be moved inside the total derivative for each m separately. [For anomalies that are not a total derivative one may replace $\exp(-\lambda^2_m/M^2)$ by $\exp(\tilde{\Box}/M^2)$ and perform the trace over a complete set of plane waves explicitly. Using the cyclicity of the trace, one may cancel the factors $e^{1/2}$ and $e^{-1/2}$ and replace $\tilde{\Box}$ by $\Box$ if there are no derivatives acting on $e^{1/2}$ and $e^{-1/2}$.]

Next we evaluate the Jacobian for the vielbein-coordinate ghost system. First we shall perform the announced direct computation, which will lead to the unwanted ∂e terms. Then we shall evaluate the Jacobian, using the factorization discussed above, and obtain the desired total derivative since no ∂e terms are encountered. Finally we shall trace the reason for the difference in both answers to the fact that the product of distributions given by $\delta(x-y)\delta'(x-y)$ can be defined in various ways.

We start from the following transformation rules:

$$\tilde{e}_\mu^m(x, \Lambda) = \tilde{e}_\mu^m(x) - \Lambda[C^\nu \partial_\nu \tilde{e}_\mu^m + (\partial_\mu C^\nu)\tilde{e}_\nu^m + k(\partial_\nu C^\nu)\tilde{e}_\mu^m] \tag{26}$$

$$\begin{aligned}\tilde{C}^m(x, \Lambda) = \tilde{C}^m(x) - \Lambda[&e_n^\nu \tilde{C}^n(\partial_\nu C^\mu)e_\mu^m \\ &+ C^\nu(\partial_\nu e_\mu^m)\tilde{C}^n e_n^\mu + (\partial_\mu C^\nu)e_\nu^m \tilde{C}^n e_n^\mu \\ &+ l\partial_\mu(C^\mu e)\tilde{C}^m e^{-1}]\end{aligned} \tag{27}$$

By partially integrating the first two ∂C terms in $\tilde{C}^m(x, \Lambda)$ we find the simpler result

$$\begin{aligned}\tilde{C}^m(x, \Lambda) = \tilde{C}^m(x) + \Lambda e^{-l}[&e_k^\mu \tilde{C}^k e_n^\nu \tilde{C}^n \partial_\nu e_\mu^m \\ &+ le^{l-1}\tilde{C}^m \partial_\mu(\tilde{C}^n e_n^\mu e^{1-l})]\end{aligned} \tag{28}$$

In (28) the ghost variation has canceled the nonscalar part of the vielbein law, leaving us with only the transport term of the vielbein and the full law for the vielbein determinant.

Computing the Jacobian in the left-hand side of (15) we get three sets of terms:

1. Terms from $\tilde{e}_\mu^m(x, \Lambda)$ writing it as a function of $\tilde{e}_\nu^m(x)$ and C^ν and keeping C^ν fixed;
2. Terms from $\tilde{e}_\mu^m(x, \Lambda)$ replacing C^ν by $\tilde{C}^n e_n^\nu e^{-l}$ and varying $e_n^\nu e^{-l} = \tilde{e}_n^\nu \tilde{e}^p$ with $p = (k-l)/(kn+1)$;
3. Terms from $\tilde{C}^m(x, \Lambda)$ keeping $e_\mu^m(x)$ fixed.

The result reads

$$J = \operatorname{sdet} \frac{\partial\{e_\mu^m(x, \Lambda), \tilde{C}^m(x, \Lambda)\}}{\gamma\{e_\nu^n(y), \tilde{C}^n(y)\}} = 1 - \Lambda \int dx(j_1 + j_2 + j_3) \tag{29}$$

For j_1 we find easily

$$j_1 = n[nC^\nu \partial_\nu + (1 + kn)(\partial_\nu C^\nu)]\delta(x-y)|_{x=y} \tag{30}$$

For j_2 we use the intermediate result

$$\frac{\partial C^\rho(x)}{\partial \tilde{e}_\nu^n(y)} \equiv \Delta_n^{\rho\nu}(x, y) = e^{-k}(pe_n^\nu C^\rho - e_n^\rho C^\nu)\delta(x-y) \tag{31}$$

and find that

$$j_2 = \sum_{\substack{m=n \\ \mu=\nu}} \Delta_n^{\rho\nu}(\partial_\rho \tilde{e}_\mu^m) + \tilde{e}_\rho^m \partial_\mu \Delta_n^{\rho\nu} + k\tilde{e}_\mu^m \partial_\rho \Delta_n^{\rho\nu}$$

$$= \{ -(l+n)(\partial_\rho C^\rho + C^\rho \partial_\rho) + (k-p-1)(\partial_m e_\rho^m) C^\rho + [1 - l + k(n+l-p)] C^\rho \partial_\rho \ln e\} \delta(x-y)|_{x=y} \tag{32}$$

Finally we find for j_3

$$j_3 = \{[1 + nl + l(l-1)] C^\rho (\partial_\rho \ln e) + (l-1) C^\rho (\partial_m e_\rho^m) + nl(\partial_\rho C^\rho) - l C^\rho \partial_\rho\} \delta(x-y)|_{x=y} \tag{33}$$

The terms without vielbein derivatives are a total derivative if

$$kn + l = \tfrac{1}{2}(n-1) \tag{34}$$

Combined with the result in (18) for the vielbein we find for the parameters k and l defined in (16) the following results:

$$k = (n-2)/2n, \qquad l = \tfrac{1}{2} \tag{35}$$

We see that the measure of the flat ghost is the same as that of a scalar. However, the terms with $\partial_\rho \ln e$ in (29) do not cancel separately and cannot be combined with the $(\partial_\rho C^\rho)$ and $C^\rho \partial_\rho$ terms into a total derivative. Nor can the $(\partial_m e_\rho^m)$ terms be handled in this way. These ∂e terms are the extra terms we mentioned above, which seem to indicate that there is a genuine coordinate anomaly.

Next we compute the Jacobian in (19) by direct means. Starting from the definition $\tilde{C}^m = e^l e_\mu^m C^\mu$ and (14), the first factor in (19) yields

$$\operatorname{sdet}[e^l(x,\Lambda) e_\nu^m(x,\Delta)\{\delta_\mu^\nu + \delta_\mu^\nu C^\rho(x) \Lambda \partial_\rho^x + \Lambda[\partial_\mu C^\nu(x)]\}\delta(x-y)] \tag{36}$$

The second factor in (19) yields the simpler result

$$\operatorname{sdet}[e^{-l}(y) e_n^\mu(y) \delta(y-z)] \tag{37}$$

We expand the vielbeins in (36) in terms of Λ and find that the two $\partial_\mu C^\nu$ terms cancel:

$$\operatorname{sdet}[\{e^l(x) e_\mu^m(x) - \Lambda(e^l(x) e_\mu^m(x) C^\nu(x) \partial_\nu^x + l e^{l-1}(x) e_\mu^m(x) \partial_\nu^x [C^\nu(x) e(x)] + e^l(x) C^\nu(x)[\partial_\nu e_\mu^m(x)])\}\delta(x-y)] \tag{38}$$

We can rewrite this result as

$$\operatorname{sdet}[e^l e_\mu^m \delta(x-y) - \Lambda\{e^l e_\mu^m C^\nu \partial_\nu + l e^l e_\mu^m (\partial_\nu C^\nu) + C^\nu[\partial_\nu(e^l e_\mu^m)]\}\delta(x-y)] \tag{39}$$

When we contract (39) with (37) we must take the trace and integrate over y. Moreover, we can bring (37) past the free derivative in (39) such that it hits the $\delta(x-y)$; subsequent integration over y then gives $\partial_x[e^{-l}(z)e^{\mu}_n(z)\delta(x-z)]$. So we arrive at

$$\begin{aligned}\mathrm{sdet}[\delta^m_m\delta(x-z) - \Lambda\{e^l(x)e^m_\mu(x)C^\nu(x)\partial^x_\nu[e^{-l}(z)e^\mu_n(z)\delta(x-z)] \\ + l\delta^n_m[\partial_\nu C^\nu(x)]\delta(x-z) + e^{-l}(x)e^\mu_n(x)C^\nu(x) \\ \times\{\partial^x_\nu[e^l(x)e^m_\mu(x)]\}\delta(x-z)]\end{aligned} \quad (40)$$

We now note that in the sum of the two terms with $C^\nu\partial_\nu$, *all* ∂e *terms cancel.* One is left with

$$\mathrm{sdet}[\delta^m_n\delta(x-y) - \Lambda\delta^m_n\{C^\nu(x)[\partial^x_\nu\delta(x-z)] + l[\partial^x_\nu C^\nu(x)]\delta(x-z)\}] \quad (41)$$

Clearly, we are in a similar situation as in the scalar case and for $l = 1/2$ the Jacobian is unity.

We have shown that two different derivations of the Jacobian give different results.

The ambiguity can be explained [18] as an ambiguity of the expression $\int dx\,dy\,f(x)\delta(x-y)\partial^x_\mu\delta(x-y)$, which we encountered in the measure of a scalar field after tracing over $x = y$. If one argues that $f(x)\delta(x-y)$ equals $f(y)\delta(x-y)$, then one can bring $f(y)$ past ∂^x_μ. One finds then

$$\int dx[f(x)\partial^x_\mu\delta(x-y)]_{x=y} = \int dx\{\partial^x_\mu[f(x)\delta(x-y)]\}_{x=y} \quad (42)$$

which would imply that $\int \partial_\mu f(x)\delta(x-y)\,dx = 0$, which is clearly incorrect.

Hence, our proposed factorization first of all means that we use the freedom due to the ambiguity to get rid of all ∂e terms and split the terms in the total Jacobian into a vielbein sector and a ghost sector. In each sector the unregularized coordinate Jacobian is a total derivative, and can be regularized by mode cutoff. (In practice one uses in the ghost sector the ghost action as regulator [5, 6], or a Hermitian function of the ghost action if the ghost action itself is not Hermitian [27, 33].)

The terms j_1 correspond to the one-but-last determinant in (15), while j_2 plus j_3 without ∂e terms correspond to (19). Note that this is different from what one might have expected naively; naively, it is j_1 plus j_2 that represents the Jacobian for the field e^m_μ.

Another way of clarifying what we have done is this. Suppose we were able to show that one could rewrite the original ill-defined expression in another way such that after regularization one would obtain a well-defined expression but with coordinate anomaly. As an act of physical prejudice, we would then select the well-defined measure that has no coordinate anomaly, instead of other well-defined measure with coordinate anomaly.

In this way one can justify the measure for the coordinate ghost and vielbein. Details can be found in Ref. 18. We shall now use these measures in applications and derive the critical dimensions of linear and nonlinear spinning strings.

3. CRITICAL DIMENSIONS OF THE SPINNING STRING

There are three classes of spinning string models, namely, the $d = 2$ supergravities with $N = 1$, $N = 2$, or $N = 4$ local supersymmetries.

The $N = 1$ supergravity model [19] can be extended to a nonlinear sigma model [34] coupled to supergravity, and one can add a Wess-Zumino (WZ) term [35] also coupled to supergravity [20]. The fields are

$$e_\mu^m, \qquad \psi_\mu^a, \qquad \phi^I, \qquad \lambda^I, \qquad I = 0, \ldots, d-1 \tag{43}$$

One may add a supergravity auxiliary field S, and a matter auxiliary field F, although they will not contribute to the critical dimension, because they appear in the action as $(\tilde{S})^2$ and $(\tilde{F})^2$ and thus decouple from gravity. One may also take a different point of view, and consider these supergravity theories as *conformal* supergravities [36]. In that case the $N = 1$ model has no auxiliary field S (but F is still present).

The $N = 2$ model has as matter fields d complex scalars ϕ^I and d complex spinors λ^I. The supergravity fields are e_μ^m, a complex ψ_μ^a, and a real vector field B_μ which gauges the group $U(1)$ [21]. It describes two ordinary spinning strings coupled to each other by B_μ [22]. One can extend it to a nonlinear sigma model with a WZ term coupled to supergravity [23]. The rigid $N = 2$ nonlinear sigma model with a WZ term was given, both in x space and in superspace, in Ref. 24. The conformal approach to the $N = 2$ model introduces extra vector gauge fields that are not present in the action but play a role in the algebra [36].

The $N = 4$ model as a nonlinear sigma model without WZ term coupled to supergravity was recently constructed [25]. This supergravity model can be extended to contain a WZ term [23]. This is then the most general sigma model with WZ term. A surprising result found in Ref. 25 is that in the rigid case one has always a hyper-Kähler geometry, but in the local case not only a quaternionic geometry but also a hyper-Kähler geometry is possible. However, it seems that with WZ term only a hyper-Kähler geometry is possible [23]. The fields in this model are e_μ^m, four $\psi_\mu^{a,I}$, a nonpropagating $SU(2)$ gauge field $\mathbf{B}_\mu$, together with $4n$ scalars ϕ^I and $4n$ Majorana spinors λ^I.

Consider the simplest model, the $N = 1$ linear sigma model without WZ term. Choosing the usual superconformal gauge for reparametrizations, local Lorentz symmetry and local supersymmetry

$$F_\alpha = \{\tilde{e}_{11} - \tilde{e}_{00} = 0, \qquad \tilde{e}_{10} + \tilde{e}_{01} = 0, \qquad \tilde{e}_{10} - \tilde{e}_{01} = 0,$$
$$\gamma_1\psi_1 - \gamma_0\psi_0 = 0, \qquad \text{all indices flat}\} \tag{44}$$

the vielbein is given by $e_\mu^m = \delta_\mu^m\sqrt{\rho}$ and the gravitino by $\psi_\mu = \gamma_\mu\phi$. (As explained in Section 2, the tildes on vielbein and gravitino are redundant in $d = 2$ dimensions.) The complete quantum action becomes

$$\begin{aligned}
\mathscr{L}(\text{quantum}) &= \mathscr{L}(\text{class}) + \mathscr{L}(\text{fix}) + \mathscr{L}(\text{ghost})\\
\mathscr{L}(\text{class}) &= -\tfrac{1}{2}(\partial_\mu\phi\partial_\nu\phi\eta^{\mu\nu}) - \tfrac{1}{2}(\bar{\lambda}\rho^{1/4})\not{\partial}_0(\rho^{1/4}\lambda)\\
\mathscr{L}(\text{fix}) &= d^\alpha F_\alpha + \text{“extra terms'' times } F_\alpha\\
\mathscr{L}(\text{ghost}) &= -i(\mathbf{C}^*\sqrt{\rho})(\tau_3\partial_1 - \gamma_0\partial_0)\mathbf{C}\\
&\quad -i(\bar{\Sigma}\rho^{-1/4})(\gamma_1\partial_1 + \gamma_0\partial_0)(\rho^{-1/4}S)\\
&\quad + i(C^*{}_{10}\sqrt{\rho})(2C_{10} + \tfrac{1}{2}(\bar{\psi}_1\gamma_0 - \bar{\psi}_0\gamma_1)S + \partial_0 C^1 + \partial_1 C^0)
\end{aligned} \tag{45}$$

The “extra terms” are due to inserting into $\mathscr{L}$(class) and $\mathscr{L}$(ghost) the gauge conditions: these extra terms are clearly proportional to F_α and only shift d^α. The coordinate ghosts C^μ have been written as vectors **C** and the bar on the supersymmetry field Σ is the Dirac bar. A derivation of this $\mathscr{L}$(quantum) was first given in Ref. 6. For an extended set of lectures on this derivation and the BRST formalism for strings, see Ref. 26.

We next replace all fields in $\mathscr{L}$(quantum) by tilded fields times the corresponding powers of det $e = \rho$. Then we remove the total ρ dependence by rescaling each tilded field. This rescaling leads to Jacobians whose product equals unity in the critical dimension.

Consider first ϕ. Its action is given by

$$-\tfrac{1}{2}\partial_\mu(\tilde{\phi}e^{-\sigma})\partial_\nu(\tilde{\phi}e^{-\sigma})\eta^{\mu\nu}, \qquad \rho \equiv e^{2\sigma} \tag{46}$$

since det $e = \rho$. We scale $\tilde{\phi}$ from $\tilde{\phi}$ to $\tilde{\phi}e^{\sigma}$ but in little steps. Suppose we have reached a point where we have det $e_\mu^m = e^{2\sigma_t} \equiv e^{2\sigma(1-t)}$. The next rescaling is thus $\tilde{\phi} \to e^{\sigma\, dt}\tilde{\phi} = (1 + \sigma\, dt)\tilde{\phi}$. As regulator we use

$$R = e^{-\sigma_t}\Box_0 e^{-\sigma_t}, \qquad \Box_0 = \partial_\mu\partial_\nu\eta^{\mu\nu} \tag{47}$$

which is Hermitian and positive definite (in Euclidean space). The Jacobian is then

$$J = 1 + \int d^2x\,\sigma(x)\,dt \int \frac{d^2k}{(2\pi)^2}\, e^{-ikx} e^{R/M^2} e^{ikx} \tag{48}$$

Pulling the place wave exp ikx to the left results in replacing ∂_μ by $\partial_\mu - ik_\mu$. The relevant k integrals all are covered by the following formula:

$$\int d^2k(2\pi)^{-2}e^{-ikx}\exp(e^{-2\alpha\sigma_t}\not\partial e^{2\beta\sigma_t}\not\partial e^{-2\alpha\sigma_t})e^{ikx}$$

$$= (4\pi)^{-1}[M^2 e^{(4\alpha-2\beta)\sigma_t} - \tfrac{1}{3}(\alpha+\beta)\Box 2\sigma_t + O(M^{-1})] \tag{49}$$

For example, for the scalars $\tilde{S}$ one has $\beta = 0$ and $\alpha = -\frac{1}{2}$.

The final Jacobian for a given field is then

$$J = \exp(cI_0), \qquad I_0 = \frac{1}{12\pi}\int d^2x\, \sigma\Box_0\sigma \tag{50}$$

where c is a constant depending on the number of field components, their statistics, and their weight of rescaling.

The coefficients c in front of the Jacobians for the various fields can be read off from $\mathscr{L}$(quantum). The results are shown in Table I. We can now at once determine the critical dimensions from the field content of the various models.

For the bosonic string ($N = 0$) the critical dimension is thus given by

$$d\cdot\tfrac{1}{2} + d\cdot\tfrac{1}{4} - 8 - 5 = 0 \Rightarrow d = 26 \tag{51}$$

For the $N = 1$ linear σ model one has

$$d\cdot\tfrac{1}{2} + d\cdot\tfrac{1}{4} - 8 - 5 + \tfrac{15}{4} + \tfrac{7}{4} = 0 \Rightarrow d = 10 \tag{52}$$

For the $N = 2$ linear σ model one has

$$2\cdot d\cdot\tfrac{1}{2} + 2\cdot d\cdot\tfrac{1}{4} - 8 - 5 + 2\cdot\tfrac{15}{4} + 2\cdot\tfrac{7}{4} - 1 = 0 \Rightarrow d = 2 \tag{53}$$

The nonpropagating $SO(2)$ field B_μ does not contribute; see the next section.

For the $N = 4$ linear σ model one has

$$4\cdot d\cdot\tfrac{1}{2} + 4\cdot d\cdot\tfrac{1}{4} - 8 - 5 + 4\cdot\tfrac{15}{4} + 4\cdot\tfrac{7}{4} - 3 = 0 \Rightarrow d = -2 \tag{54}$$

Table I. Conformal Anomalies for the Various Fields

Field	Value of c	
One real scalar field	$\frac{1}{2}$	Ref. 4
Two real coordinate ghosts C^μ	-8	
Two real coordinate antighosts $C^{*\mu}$	-5	
A Majorana fermion λ	$\frac{1}{4}$	Ref. 6
Two anti-Majorana[a] susy ghosts S^a	$\frac{15}{4}$	
Two Majorana susy antighosts Σ^a	$\frac{7}{4}$	
One real scalar Maxwell ghost	-1	Ref. 27
One real scalar Maxwell antighost	0	

[a] The coordinate ghosts and supersymmetry ghosts always have opposite reality properties; see Ref. 26.

The nonpropagating $SU(2)$ gauge field B_μ does not contribute; see the next section.

Since in two dimensions there are no transversal modes and negative dimensions have no physical significance, only the cases $d = 26$ and $d = 10$ with $N = 0$ and $N = 1$, respectively, remain.

The divergent M^2 terms cancel in supersymmetric models. Their contribution is equal to the sum of the numbers of bosonic field components minus the number of fermionic field components, each weighted with the weight of the rescaling [6]. This sum cancels [22], but note that this sum is not the same sum as just the difference of the number of bosonic and fermionic field components [6].

For the $N = 2$ vector fields B_μ, the Abelian $U(1)$ gauge invariance is fixed by the gauge fixing term $\partial_\mu B_\nu \eta^{\mu\nu} = 0$. Note that in $d = 2$, B_μ equals $\tilde{B}_\mu$ so that the gauge condition is linear in tilded fields [6]. The Maxwell ghost action becomes

$$C^*(\partial_\mu \partial_\nu \eta^{\mu\nu})(\det e^{-1/2} \tilde{C}) \tag{55}$$

where C^* and $C = \tilde{C} \det e^{-1/2}$ are the Maxwell antighost and ghost. As *Hermitian* regulator for the Maxwell ghost we choose

$$R = (\det e^{-1/2}) \Box_0 \Box_0 (\det e^{-1/2}) \tag{56}$$

Note that we must rescale $\tilde{C}$ by $\tilde{C} \to \tilde{C} e^{+\sigma}$ but C^* is not rescaled. Doing the k integral, one finds $c = -1$ for the ghost and $c = 0$ for the antighost. In an earlier study [6], a Hermitian ghost action $C^*(\det e)^{-1/2}(\det e)^{-1/2}\tilde{C}$ was obtained which yielded $c = -\frac{1}{2}$ and $c = -\frac{1}{2}$ instead of $c = -1$ and $c = 0$. A detailed analysis [27] shows that indeed for Maxwell ghosts the c for $\Box_0\Box_0$ is twice the c for $\Box_0$, but we consider the argument used to obtain the Hermitian ghost action with $\Box_0$ instead of $\Box_0^2$ no longer convincing.

4. NONLINEAR σ MODELS

The critical dimensions for nonlinear σ models are not 26 (for the bosonic case) and 10 (for the supersymmetric case) but lower. In fact, if space-time splits up into a direct product of a d-dimensional Minkowski space-time with coordinates $X^I(\sigma, \tau)$ $(I = 0, \ldots, d-1)$ and a group manifold G with coordinates $\phi^i(\sigma, \tau)$ $(i = 1, \ldots, \dim G)$, then the critical dimension $d = d(\text{crit})$ is given by [42]

$$\begin{aligned} d(\text{crit}) &= 26 - (\dim G)/\gamma \\ d(\text{crit}) &= 10 - \tfrac{1}{3}\dim G - \tfrac{2}{3}\dim G/\gamma \\ \gamma &= 1 + \tfrac{1}{2} C_2(G) k^{-1} \end{aligned} \tag{57}$$

The symbol $C_2(G)$ is the second Casimir operator with a particular normalization [for $SO(N)$, it is given by $2(N-2)$], and k is the overall constant of the action.

The action for the most general $N = 1$ locally supersymmetric nonlinear σ model consists of four kinds of terms [20]:

i. The rigidly supersymmetric nonlinear σ model;
ii. The (two) Noether terms for the locally supersymmetric linear σ model;
iii. The rigidly supersymmetric Wess-Zumino terms;
iv. One extra term needed for locally supersymmetrizing the WZ term.

In particular, the two terms in (ii) are by themselves already sufficient for the locally supersymmetric nonlinear σ model without WZ term.

In addition to this $N = 1$ model, there also exist locally supersymmetric nonlinear σ models (with or without WZ term) with $N = 2$ and $N = 4$ supersymmetries. We refer to the previous section.

Let us now take a closer look at the $N = 1$ model. The action reads

$$
\begin{aligned}
\mathscr{L}(\text{nonlin}) &= (k/32\pi^2)[\mathscr{L}(\text{i}) + \mathscr{L}(\text{ii}) + \mathscr{L}(\text{iii}) + \mathscr{L}(\text{iv})] \\
\mathscr{L}(\text{i}) &= -\tfrac{1}{2}(\partial_\mu\phi^i\partial_\nu\phi^j)eg^{\mu\nu}g_{ij}(\phi) \\
&\quad -\frac{e}{2}\bar{\chi}^i\gamma^\mu(\partial_\mu\chi^j + \{{}^{\,j}_{kl}\}\partial_\mu\phi^k\chi^l)g_{ij}(\phi) \\
&\quad -\frac{e}{12}R_{ijkl}(\bar{\chi}^i\chi^k)(\bar{\chi}^j\chi^l) \\
\mathscr{L}(\text{ii}) &= \text{a term } (\bar{\psi}_\mu\gamma^\nu\gamma^\mu\chi^i)\partial_\nu\phi^j g_{ij} \\
&\quad\ \text{and a term with } (\bar{\psi}_\mu\gamma^{\mu\nu}\psi_\nu)[\bar{\chi}^i\chi^j g_{ij}(\phi)] \\
\mathscr{L}(\text{iii}) &= -\tfrac{1}{2}(\partial_\mu\phi^i\partial_\nu\phi^j)\varepsilon^{\mu\nu}B_{ij}(\phi) \\
&\quad + F_{ijk}(\bar{\chi}^i\gamma^\mu\tau_3\chi^j)\partial_\mu\phi^k \\
&\quad + \text{more terms with } \chi \text{ times } F \\
\mathscr{L}(\text{iv}) &= \text{a term with } F\psi_\mu\chi^3
\end{aligned}
\tag{58}
$$

The symbol F_{ijk} is the curl of B_{ij}.

This action can be written down for arbitrary metric $g_{ij}(\phi)$ on a trivial manifold, in which case k is arbitrary. However, in order that $B_{ij}(\phi)$ be able to be defined consistently in different patches of nontrivial manifolds on which g_{ij} is the metric, the coefficient of the bosonic WZ term has to be quantized: k must be a (positive or negative) integer while F_{ijk} is taken proportional to the structure constants of the group G (see below). Furthermore, in order that the model has vanishing $\Delta\mathscr{L}$(1-loop) in *flat space*, which implies a vanishing one-loop β function, which implies (in this case, not in general) vanishing T^{α}_{α} of the one-loop effective action, the relative strength between the non-WZ-terms and the WZ-terms is fixed. Thus the string tension [the coefficient in front of $(\partial_\mu \phi)^2$] is related to the size of the compact manifold (the coefficient in front of the WZ term). Finally, in order that also in curved space (i.e., with world-sheet metric $g_{\mu\nu} \neq \eta_{\mu\nu}$) $\Delta\mathscr{L}$(1-loop) vanishes, one must select $d = d(\text{crit})$, and that aspect we will discuss in this section. In that case the model is locally Weyl invariant. (Local Weyl invariance in curved space is equivalent to conformal invariance in flat space; local conformal invariance in curved space is discussed in Ref. 30.)

We thus consider $\mathscr{L}$(nonlin) on a group manifold, parametrized by ϕ^i, with the matrices $G = G(\phi)$ forming a representation of the group G. (Which representation does not matter, as long as it is faithful.) We define group left-vielbeins by

$$G^{-1}\partial_\mu G = e_i^a(\phi) T_a \partial_\mu \phi^i \tag{59}$$

where T_a are the generators of G and $\partial_\mu = \{\partial_\sigma, \partial_\tau\}$, and a group right-vielbeins by

$$\partial_\mu G G^{-1} = f_i^a(\phi) T_a \partial_\mu \phi^i \tag{60}$$

The group metric is then given by

$$g_{ij}(\phi) = e_i^a(\phi) e_j^b(\phi) \gamma_{ab} = f_i^a(\phi) f_j^b(\phi) \gamma_{ab} \tag{61}$$

where γ_{ab} is the Killing metric. (Since the Killing metric is constructed from the structure constants and the latter are invariant tensors of the adjoint representation $A_b^a = e_i^a f_b^i$, one can express the metric either in terms of e_i^a or f_i^a.) The curl of B_{ij} is related to the structure constants as

$$e_a^i e_b^j e_c^k F_{ijk} = f_{abc} = f_a^i f_b^j f_c^k F_{ijk} \tag{62}$$

Summarizing, at the one-loop level, using a background field computation

$$eT^{\alpha}_{\alpha} = (\beta_{ij}^G e g^{\mu\nu} + \beta_{ij}^B \varepsilon^{\mu\nu})(\partial_\mu \phi^i \partial_\nu \phi^j) + \beta^\Phi e R^{(2)} \tag{63}$$

In flat space (and in curved space) $\beta_{ij}^G = 0$, and $\beta_{ij}^B = 0$ if k is quantized and $F_{abc} = f_{abc}$, while $\beta^\Phi = 0$ if $d = d(\text{crit})$.

With this choice for F_{ijk} all χ^4 terms cancel from the action, and the F_{ijk} terms can be interpreted as torsion that parallelizes the group manifold [31]. ("Parallelization" means that the full Riemann curvature built from the spin connection plus torsion, vanishes.) Adopting the superconformal gauge.

$$e_\mu^m(x) = e^{\sigma(x)} \delta_\mu^m, \qquad \psi_\mu(x) = \gamma_\mu \eta \tag{64}$$

all gravitino-dependent terms cancel from the action. The action reduces in this case to a very simple form

$$\begin{aligned}\mathscr{L}(\text{nonlin}) = &-\tfrac{1}{2}(\partial_\mu \phi^i \partial_\nu \phi^j)(\delta^{\mu\nu} g_{ij} + \varepsilon^{\mu\nu} B_{ij}) \\ &- (e/2)(\bar{\chi}^i_L e^a_i) \partial\!\!\!/_0 (e^b_j \chi^j_L)\gamma_{ab} \\ &- (e/2)(\bar{\chi}^i_R f^a_i) \partial\!\!\!/ (\chi^j_R f^b_j)\gamma_{ab}\end{aligned} \tag{65}$$

with $\partial\!\!\!/_0 = \gamma^m \delta^\mu_m \gamma_\mu$ and $\chi^i_L = \frac{1}{2}(1+\tau_3)\chi^i$ while $\chi^i_R = \frac{1}{2}(1-\tau_3)\chi^i$.

The fermions $\chi^a_L \equiv \chi^i_L e^a_i$ and $\chi^a_R = \chi^i_R f^a_i$ are thus decoupled from the scalars and only couple to gravity, but the scalars are still horribly nonlinear. Taking for the flat fermions as measure [27]

$$D(\chi^a_L e^{1/2}) D(\chi^a_R e^{1/2}) \tag{66}$$

we are in the same situation as for linear σ models, as far as the fermions are concerned. By rescaling the linear σ model in space-time as before, and also these internal fermions, we find as criterion for the Jacobian to vanish

$$\tfrac{1}{2}d + \tfrac{1}{4}d - 13 + \tfrac{11}{2} + \tfrac{1}{4}\dim G + \text{contribution of internal scalars} = 0 \tag{67}$$

or,

$$d(\text{crit}) = 10 - \tfrac{1}{3}\dim G - \tfrac{4}{3}(\text{contribution of internal scalars}) \tag{68}$$

The origin of the term $-\frac{1}{3}\dim G$ in (57) is now clear: it is due to the free internal fermions of which there are as many as dim G. We must now try to decouple the scalars from gravity.

In the superconformal gauge, $g^{\mu\nu}\sqrt{g} = \eta^{\mu\nu}$, but replacing $\phi^i = \tilde{\phi}^i e^{-1/2}$, the rescalings of $\tilde{\phi}^i$ that are to remove the vielbein dependence of the ϕ action must be regularized by the action which is very nonlinear. Consequently it is very difficult to perform these non-Gaussian integrals. Instead, we will use non-Abelian bosonization formulas which equate the nonlinear bosonic path integral for ϕ^i to a much simpler fermionic path integral. The identity we shall assume and which is discussed in [32] reads

$$\begin{aligned}&Z_{\text{bos}}(\phi^i, g_{\mu\nu}, k, SO(N)) \\ &\quad = \int d\tilde{A}_\mu \, d\tilde{\lambda}^I \exp\left[-\sum_I \int d^2\chi \, e \bar{\lambda}^I_a \gamma^\mu (\partial_\mu \delta^{ab} + A^{ab}_\mu)\lambda^I_b\right]\end{aligned} \tag{69}$$

where $A^{ab}_\mu = A_\mu \cdot T^{ab}$ is an $SO(k)$ gauge field, and the N spinors λ^I $(I = 1, N)$ are in the fundamental representation of $SO(k)$ $(a, b = 1, k)$. On the two-dimensional world sheet, $\tilde{A}_\mu = A_\mu$ but $\tilde{\lambda}^I = \lambda^I e^{1/2}$. The right-hand side depends only on e^m_μ, just like the left-hand side, but the exponent is only quadratic and cubic in fields. We shall now perform the actual rescalings which decouple gravity (e^m_μ) from the matter fields ($\tilde{\lambda}^I_a$ and $\tilde{A}_\mu$). The infinitesimal Jacobians we will sum and in the critical dimension the sum must vanish.

We begin with a useful simple lemma [37].

Lemma. For any two-dimensional vector field A_μ we can write

$$\partial\!\!\!/ + A\!\!\!/ = h^{\dagger -1} \partial\!\!\!/ h^{-1}, \qquad h = \begin{pmatrix} h_+ & 0 \\ 0 & h_- \end{pmatrix} = \exp(i\eta + \tau_3\phi)$$

$$h_+^{\dagger -1} = h_-, \qquad h^{\dagger -1} = \exp(i\eta - \tau_3\phi) \tag{70}$$

where A is Lie-algebra-valued and anti-Hermitian, while η and ϕ are Lie-algebra-valued and Hermitian.

Proof. The proof is simple in the representation

$$\gamma_1 = \begin{pmatrix} 0 & 1 \\ 1 & 0 \end{pmatrix}, \qquad \gamma_2 = \begin{pmatrix} 0 & -i \\ i & 0 \end{pmatrix}, \qquad \tau_3 = \begin{pmatrix} 1 & 0 \\ 0 & -1 \end{pmatrix} \tag{71}$$

We have

$$\partial\!\!\!/ + A\!\!\!/ = \begin{pmatrix} 0 & \partial_- + A_- \\ \partial_+ + A_+ & 0 \end{pmatrix}$$

so we must solve

$$A_- = h_+^{\dagger -1}(\partial_- h_-^{-1}) = h_-(\partial_- h_-^{-1}) = -(\partial_- h_-)h_-^{-1}$$

where we used $h_+^{\dagger -1} = h_-$, something that defines h_+ in terms of h_-. The equation $\partial_- h_- = -A_- h_-$ is a first-order equation for h_- for given A_- and has as solution

$$h_- = \left[P \exp \int_0^{x^-} - A_-(x'_-, x_+)\, dx'_- \right] A_-(x_- = 0, x_+) \tag{72}$$

With h_- given, one may then verify that the equation for A_+ is satisfied as well:

$$A_+ = h_-^{\dagger -1} \partial_+ h_+^{-1} = h_+ \partial_+ h_+^{-1} \tag{73}$$

because under Hermitian conjugation it goes over into the equation for A_-:

$$(A_+)^\dagger = -A_- = (\partial_- h_+^{-1\dagger}) h_+^\dagger = (\partial_- h_-) h_-^{-1}$$

We can thus rewrite the action in the superconformal gauge and on the basis $\tilde{\lambda}^I$ as follows:

$$I = \int d^2x - \sum_{I=1}^{N} (\bar{\tilde{\lambda}}^I e^{-\sigma/2} h^{\dagger -1}) \partial\!\!\!/_0 (h^{-1} e^{-\sigma/2} \tilde{\lambda}^I) \tag{74}$$

where $\partial\!\!\!/_0 = \gamma_m \partial_\nu \eta^{m\nu}$ is constant. A small rescaling of $\tilde{\lambda}^I$ which removes a bit of $e^{-\sigma/2}$ and of h^{-1}, proportional to dt, yields as Jacobian

$$\text{“tr”}\; dt\left(\frac{\sigma}{2} + i\eta + \tau_3\phi\right) \tag{75}$$

where the trace includes an integral over space-time. To regularize this expression we use as regulator the square of the action, and must compute

$$K = \int \frac{d^2k}{(2\pi)^2} e^{-ikx} \exp\left(\frac{R}{M^2}\right) e^{ikx}, \qquad M^2 \to \infty$$
$$R = e^{-2\alpha\sigma} \not{D} e^{2\beta\sigma} \not{D} e^{-2\alpha\sigma}, \qquad \not{D} = \not{\partial}_0 + \not{A} \tag{76}$$

where $\not{A} = \gamma_m A_\mu \delta^{m\mu}$ is again independent of gravity and $\alpha = \frac{1}{4}$, $\beta = -\frac{1}{2}$. The computation of this integral is not hard, but for completeness it is given at the end of this section in a separate appendix. One finds for the integral

$$K = \frac{-1}{12\pi}[-3M^2 e^{(4\alpha-2\beta)\sigma} - \tfrac{3}{2}\gamma^{mn}\delta^\mu_m \delta^\nu_n F_{\mu\nu} + 2(\alpha+\beta)\Box_0\sigma + \mathcal{O}M^{-2}] \tag{77}$$

The term with M^2 is divergent and is usually dropped on somewhat unclear grounds. [In the zeta-function regularization, one begins by adding on top of the regulator $\exp(-\lambda_m \tau)$ a second regulator in the form of $\Gamma(s)^{-1} t^{s-1}$. Afterwards one then finds that owing to this second regulator the M^2 terms vanish in the limit $s \to 0$. However, this means only that by choosing the second regulator one has thrown the divergence away, just as choosing dimensional regularization is equivalent to only retaining logarithmically divergent integrals.]

Suppose we have already decreased $h = \exp(i\eta + \tau_3\phi)$ to $h_t \equiv \exp(1-t)(i\eta + \tau_3\phi)$ and σ to $\sigma_t \equiv (1-t)\sigma$. ($F^t_{\mu\nu}$ depends on A^t_μ, which is constructed from h_t as before.) Then the infinitesimal Jacobian reads

$$+\frac{N}{12\pi}\operatorname{Tr}\int d^2x \int_0^1 dt\left(\frac{\sigma}{2} + i\eta + \tau_3\phi\right)\left(-\frac{3}{2}\gamma_0^{\mu\nu}F^t_{\mu\nu} - \tfrac{1}{2}\Box_0\sigma_t\right) \tag{78}$$

(note the minus sign for the fermionic trace) where Tr denotes a trace over Dirac and group indices. The Jacobian thus splits into a purely gravitational part

$$-N/(48\pi)\operatorname{Tr}\int d^2x \int_0^1 dt\, \sigma\Box_0\sigma_t$$
$$= +\tfrac{1}{4}NkI_0 \qquad \text{[see equation (50)]} \tag{79}$$

and a purely Yang–Mills part

$$-N/(8\pi)\operatorname{Tr}\int d^2x \int_0^1 dt\, \tau_3\phi\gamma_0^{\mu\nu}F^t_{\mu\nu}$$
$$= -N/(4\pi)\operatorname{Tr}\int d^2x \int_0^1 dt\, \tau_3\phi(\not{\partial}_0\not{A}^t + \not{A}^t\not{A}^t) \tag{80}$$

Terms with η cancel, because the generators of $SO(N)$ in the vector representation are traceless, while $\gamma^{\mu\nu}$ is traceless as well, and cross terms between the gravity and Yang-Mills sector cancel for the same reason.

The Yang-Mills Jacobian can be expressed in a more manageable form by another lemma:

Lemma.

$$\frac{1}{2}\frac{d}{dt}\operatorname{Tr} A\!\!\!/^t A\!\!\!/^t = -2\operatorname{Tr}\tau_3 A\!\!\!/^t \phi A\!\!\!/^t + 2i\operatorname{tr} A^t_\mu \partial^\mu \eta - \operatorname{Tr}\tau_3(\partial\!\!\!/_0\phi)A\!\!\!/^t \tag{81}$$

Proof. Recalling the definitions

$$A\!\!\!/^t = h_t^{-1\dagger}\partial\!\!\!/_0 h_t^{-1}, \qquad h_t = \exp(1-t)(i\eta + \tau_3\phi) \tag{82}$$

we obtain

$$\frac{d}{dt}A\!\!\!/^t = \frac{d}{dt}[e^{(t-1)(-i\eta+\tau_3\phi)}\partial\!\!\!/_0 e^{(t-1)(i\eta+\tau_3\phi)}] = (-i\eta+\tau_3\phi)A\!\!\!/^t + A\!\!\!/_t(i\eta+\tau_3\phi) + \partial\!\!\!/_0(i\eta+\tau_3\phi) \tag{83}$$

Owing to the cyclicity of the trace we have

$$\frac{1}{2}\frac{d}{dt}\operatorname{Tr} A\!\!\!/^t A\!\!\!/^t = \operatorname{Tr} A\!\!\!/^t \frac{d}{dt} A\!\!\!/^t \tag{84}$$

which proves the lemma.

Using this lemma, we eliminate $\operatorname{Tr}\tau_3\phi\partial\!\!\!/ A^t = -\operatorname{Tr}\tau_3(\partial\!\!\!/\phi)A^t$ and obtain

$$J(\text{Yang-Mills}) = \exp\left\{(-N/4\pi)\int_0^1 dt \int d^2x \times \operatorname{Tr}\left[\tau_3 A\!\!\!/^t\phi A\!\!\!/^t - iA\!\!\!/^t\partial\!\!\!/_0\eta + \frac{1}{2}\frac{d}{dt}(A\!\!\!/^t A\!\!\!/^t)\right]\right\} \tag{85}$$

or

$$J(\text{Yang-Mills}) = \exp\left\{(N/4\pi)\int d^2x \times\left[\operatorname{tr}_c(A_\mu A_\nu \delta^{\mu\nu}) + \operatorname{tr}_c\left(\int_0^1 dt\, 2iA^t_\mu \partial_\nu \eta\delta^{\mu\nu}\right) - \int_0^1 dt\operatorname{Tr}(\tau_3 A\!\!\!/^t\phi A\!\!\!/^t)\right]\right\} \tag{86}$$

where tr_c denotes a trace over groups indices only.

Thus, we have achieved the decoupling of gravity and the $SO(k)$ vector field A_μ from the fermions $\tilde{\lambda}^I$. The Jacobian consists of a purely gravitational part

$$+\tfrac{1}{4}kNI_0 \tag{87}$$

and a Yang-Mills part J(Yang-Mills) which is given above. At first sight one would conclude that this solves our problems since we decoupled gravity from the rest (from $\tilde{\lambda}^I$ and A_μ). However, η and ϕ depend on A_μ *nonlocally*. For example, for $k = 2$ the $SO(2)$ generator commutes with itself, and A becomes equal to

$$\not{A} = \exp(i\eta - \tau_3\phi)\not{\partial}_0 \exp(-i\eta - \tau_3\phi) = -\not{\partial}_0(i\eta + \tau_3\phi) \tag{88}$$

so that $A_\mu = -i(\partial_\mu\eta + \varepsilon_{\mu\nu}\partial^\nu)$ with $\varepsilon_{12} = \varepsilon^{12} = +1$. In that case the Jacobian becomes proportional to

$$A_\mu(\delta^{\mu\nu} - \partial^\mu\partial^\nu\Box_0^{-1})A_\nu \tag{89}$$

and the nonlocal operator $\Box_0^{-1}$ *is expected to feel the effects of curved space.*

We now first discuss the case $k = 2$, and later the case by $k > 2$. Let us change integration variables in the path integral from $\tilde{A}_\mu = A_\mu$ to (η, ϕ). The Jacobian is proportional to $\Box_0$, and exponentiating by means of anticommuting auxiliary fields Z ("ghosts," similar to the familiar Faddeev-Popov ghosts) we get an action

$$\bar{Z}\Box_0 Z \tag{90}$$

The integration measure of $\bar{Z}$ and Z we take as $\tilde{\bar{Z}} = \bar{Z}e^{1/2}$ and $\tilde{Z} = Ze^{1/2}$ because $\bar{Z}\Box_0 Z$ is generally covariant outside the conformal gauge if we let both $\bar{Z}$ and Z transform as scalars under BRST coordinate transformations. (Note that $\bar{Z}\Box_0 Z$ corresponds to a generally coordinate invariant action in the conformal gauge.) To remove the vielbein dependence we rescale $\tilde{\bar{Z}}$ and $\tilde{Z}$, just as we rescaled the scalars before, and find a Jacobian

$$-I_0 \tag{91}$$

(twice as much as for a real scalar because $\bar{Z}$ and Z are independent, and a minus sign since $\bar{Z}$ and Z are ghosts). In terms of η, ϕ the action becomes

$$\phi\Box_0\phi \tag{92}$$

Its independence of η reflects the original local $SO(k)$ gauge invariance.

We now will argue that one should use as integration variables $(\tilde{\eta}, \tilde{\phi})$ and not (η, ϕ). To see why, we begin by noting that the decomposition $A_\mu = -i(\partial_\mu\eta + \varepsilon_{\mu\nu}\partial^\nu\phi)$ in the conformal gauge corresponds to

$$A_\mu = -i(\partial_\mu\eta + \varepsilon_{\mu\nu}eg^{\nu\rho}\partial_\rho\phi) \tag{93}$$

in a general gauge. Hence, both η and ϕ are good coordinate scalars. In order to avoid a local coordinate anomaly we must use $\tilde{\phi}$ as integration variable. Removal of the vielbein dependence from the action

$$(\tilde{\phi}e^{-1/2})\Box_0(e^{-1/2}\tilde{\phi}) \tag{94}$$

then yields a Jacobian

$$+\tfrac{1}{2}I_0 \tag{95}$$

We fix the local $SO(k)$ gauge invariance [$\delta\eta = f(x)$ with $f(x)$ an arbitrary $SO(k)$-valued parameter] by $\eta = 0$. The ghosts are then algebraic and one gets no contribution from this sector. (Taking as integration variable $\tilde{\eta}$ does not make any difference for these nonpropagating fields.)

Adding all contributions we get for the $k = 2$ case a total Jacobian proportional to [33]

$$\tfrac{1}{2}d + \tfrac{1}{4}d - 13 + \tfrac{11}{2} + \tfrac{1}{4}\dim G + (+\tfrac{1}{4}Nk - 1 + 1/2) \tag{96}$$

in agreement with (57).

Finally we consider the case $k > 2$. We start again with the Jacobian which was the result of rescaling the fermions (due to fermionizing the scalars of the nonlinear σ model)

$$J = \exp\left\{\frac{1}{4}NkI_0 + \frac{N}{4\pi}\int d^2x\left[\operatorname{tr}_c A_\mu A_\nu \delta^{\mu\nu}\right.\right.$$
$$\left.\left. + 2i\int_0^1 dt\,\operatorname{tr}_c A_\mu^t \partial_\nu \eta \delta^{\mu\nu} - \int_0^1 dt\,\operatorname{Tr}(\tau_3 \not{A}^t \phi \not{A}^t)\right]\right\} \tag{97}$$

Now, with $k > 2$, we have a non-Abelian group, and the last term does not vanish, while also A_μ is an exponential function of η and ϕ. As a result the t integral seems rather difficult to perform directly and we proceed instead differently by choosing a convenient gauge. We choose the gauge $A_+ = 0$. In this gauge the A-dependent terms in (97) reduce to

$$i\int_0^1 dt\,\operatorname{tr}_c A_-^t \partial_+ \eta, \qquad A_- = A_1 - iA_2 \tag{98}$$

We now use the results of Ref. 37 and show that (98) is actually again a nonlinear σ model. In the gauge $A_+ = 0$ we have $h_+^t = 1$ so that $\eta + \phi = 0$. It follows that h_-^t is equal to $\exp(1-t)2i\eta \equiv U_t$. Thus $A_-^t = h_-^t \partial_-(h_-^t)^{-1}$ is equal to $U_t \partial_- U_t^{-1}$. Since $\varepsilon^{-+} = \delta^{-+} = +1$, we obtain for (98)

$$i\int_0^1 dt\,\operatorname{tr}_c(\delta^{\mu\nu} + i\varepsilon^{\mu\nu})(U_t \partial_\mu U_t^{-1})\partial_\nu \eta \tag{99}$$

Next we derive the following identities:

$$\frac{d}{dt}(\partial_\mu U_t^{-1}\partial^\mu U_t) = 4i(U_t\partial_\mu U_t^{-1})\partial^\mu\eta$$
$$\varepsilon^{\mu\nu}(U_t\partial_\mu U_t^{-1})\partial_\nu\eta = \frac{-i}{2}\varepsilon^{\mu\nu}V_\mu^t V_\nu^t V_t^t, \qquad V_\mu^t = (\partial_\mu U_t)U_t^{-1} \tag{100}$$

We used the fact that η commutes with U_t and we partially integrated. Inserting (100) into (99), we find for the Jacobian due to (98) the following result:

$$J = \exp\frac{N}{16\pi}\operatorname{tr}_c\left\{\int d^2x\, \partial_\mu U^{-1}(t=0)\partial^\mu U(t=0)\right.$$
$$\left.+\int d^3x\, 2i\varepsilon^{\mu\nu}(\partial_\mu U_t)U_t^{-1}(\partial_\nu U_t)U_t^{-1}(\partial_t U_t)U_t^{-1}\right\} \tag{101}$$

Replacing $2\varepsilon^{\mu\nu}$ by $\frac{2}{3}\varepsilon^{\mu\nu t}$, we find the correct relative strength for a bosonic nonlinear σ model with WZ term and group $SO(k)$ [35]. The overall constant is now N instead of k, but note that the overall sign is opposite to what one usually has. If we denote the total coefficient of I_0 [defined in (50)] due to (69) by $C(N, k)$, then the contribution from (101) is given by

$$-C(|k|, N) \tag{102}$$

Next we change the integration variables from A_- to η. From $\delta A_- = D_-(h_-\delta h_-^{-1})$ and $h_-\delta h_-^{-1} \sim \delta\eta$, we see that the Jacobian for this basic change is proportional to det D_-. However, Polyakov and Wiegmann [38] already showed that det $\not{D}$ = det D_+ det D_- is proportional to a nonlinear σ model with WZ terms and with group $SO(k)$ but with opposite overall sign and multiplied by an overall factor $C_2(SO(k)) = 2(k-2)$. Hence this Jacobian gives an overall contribution $(-2k+4)C(|k|, 1) = C(|k|, -2k+4)$.

The ghost action in the unweighted gauge $A_+ = 0$ reads $C^*\partial_+C$. We rescale C into $\tilde{C}e^{-1/2}$ but keep $C^* = \tilde{C}^*$ since C and C^* are standard Faddeev-Popov ghosts (so that C^* transforms into d fields under BRST). Iterating the ghost kinetic term we find as regulator $R = (\exp - \sigma_t)\partial_-\partial_+(\exp - \sigma_t) = (\exp - \sigma_t)\Box_0(\exp - \sigma_t)$ and thus these ghosts yield a contribution

$$-\tfrac{1}{2}k(k-1)I_0 \tag{103}$$

Adding the various contributions we get a recursive relation for the total Jacobian $C(N, k)I_0$ of the original nonlinear σ model, in terms of $C(k, -N-2(k-2))$. In fact, one obtains

$$C(N, k) = \tfrac{1}{4}kN - C(|k|, N+2k-4) - \tfrac{1}{2}k(k-1) \tag{104}$$

This equation can be solved, and the result agrees with the critical dimension obtained from the operator approach [33].

AKNOWLEDGMENTS. The results discussed above were obtained in collaboration with P. Bouwknecht, with M. Rocek and C. S. Zhang, with A. Eastough, L. Mezincescu, and E. Sezgin, with K. Fujikawa, U. Lindström, and N. K. Nielsen, and with A. Ceresole, A. Lerda, and P. Pizzochero. It is a pleasure to thank them for their cooperation. This research was supported in part by NSF grant No. PHY 85 07627.

REFERENCES

1. K. Fujikawa, *Phys. Rev. Lett. 42*, 1195 (1979).
2. K. Fujikawa, *Phys. Rev. D 21*, 2848 (1980); *22*, 1499. (1981), (erratum) *29*, 285 (1984).
3. K. Fujikawa, *Phys. Rev. Lett. 44*, 1733 (1980); *Phys. Lett. 108B*, 33 (1982); *Phys. Rev. D 23*, 2262 (1981).
4. A. M. Polyakov, *Phys. Lett. 103B*, 211 (1981).
5. K. Fujikawa, *Phys. Rev. D 25*, 2584 (1982).
6. P. Bouwknecht and P. van Nieuwenhuizen, *Class. Quantum Grav. 3*, 207 (1986).
7. P. H. Frampton, D. R. T. Jones, S. C. Zhang, and P. van Nieuwenhiuzen, contribution to Festschrift for E.S. Fradkin (C. Isham, ed.).
8. M. J. Duff and P. van Nieuwenhuizen, *Phys. Lett. 94B*, 179 (1980).
9. M. T. Grisaru, N. K. Nielsen, W. Siegel, and D. Zanon, *Nucl. Phys. B247*, 157 (1984).
10. K. Fujikawa, *Nucl. Phys. B226*, 437 (1984); K. Fujikawa and O. Yasuda, *Nucl. Phys. B245*, 446 (1984).
11. W. A. Bardeen and B. Zumino, *Nucl. Phys. B244*, 421 (1984).
12. S. Deser and P. van Nieuwenhuizen, *Phys. Rev. D 10*, 411 (1974).
13. P. van Nieuwenhuizen, *Phys. Rev. D 24*, 3315 (1981).
14. P. van Nieuwenhuizen, *Phys. Rep. 68*, 189, Sec. 2 (1981).
15. N. K. Nielsen, *Phys. Lett. 103B*, 197 (1981).
16. F. R. Ore and P. van Nieuwenhuizen, *Phys. Lett. 112B*, 364 (1982); *Nucl. Phys. B204*, 317 (1982).
17. P. van Nieuwenhuizen, Proceedings 1985 Scottish Universities Summer School (A. T. Davies and D. G. Sutherland, eds.).
18. K. Fujikawa, U. Lindstrom, N. K. Nielsen, M. Rocek, and P. van Nieuwenhuizen, Phys. Rev. D, to be published.
19. L. Brink, P. di Vecchia, and P. Howe, *Phys. Lett. 65B*, 471 (1976); S. Deser and B. Zumino, *Phys. Lett. 65B*, 369 (1976).
20. E. Bergshoeff, S. Randjbar-Daemi, A. Salam, H. Sarmadi, and E. Sezgin, *Nucl. Phys. B269*, 389 (1986).
21. L. Brink and J. H. Schwarz, *Nucl. Phys. B121*, 285 (1977).
22. E. S. Fradkin and A. A. Tseytlin, *Phys. Lett. 106B*, 63 (1981).
23. B. de Wit and P. van Nieuwenhuizen, to be published.
24. S. J. Gates, C. M. Hull, and M. Roček, *Nucl. Phys. B248*, 157 (1984).
25. M. Pernici and P. van Nieuwenhuizen, *Phys. Lett. 169B*, 381 (1986).
26. P. van Nieuwenhuizen, Proceedings of the 1986 Physics School at Dubrovnik (M. Martinis and I. Andrič, eds.), World Scientific, Singapore, 1987.

27. A. Eastough, L. Mezincescu, E. Sezgin, and P. van Nieuwenhuizen, *Phys. Rev. Lett. 57*, 29 (1986).
28. See Ref. 22 below equation (24).
29. M. K. Fung, D. R. T. Jones, and P. van Nieuwenhuizen, *Phys. Rev. D 22*, 2995 (1980).
30. M. Kaku, P. K. Townsend, and P. van Nieuwenhuizen, *Phys. Rev. D 17*, 3179 (1978).
31. T. L. Curtright and C. K. Zachos, *Phys. Rev. Lett. 53*, 1977 (1984).
32. I. Antoniadis and C. Bachas, *Nucl. Phys. B278*, 343 (1986); A. N. Redlich and H. J. Schnitzer, *Phys. Lett. 167B*, 315 (1986); D. Chang, A. Kumar, and R. N. Mohapatra, *Z. Phys. C 32*, 417 (1986).
33. A. Ceresole, A. Lerda, P. Pizzochero, and P. van Nieuwenhuizen, *Phys. Lett. 189B*, 34 (1987).
34. L. Alvarez Gaumé and D. Z. Freedman, *Commun. Math. Phys. 80*, 443 (1981).
35. E. Witten, *Commun. Math. Phys. 92*, 455 (1983).
36. P. van Nieuwenhuizen, *Int. J. Mod. Phys. A 1*, 155 (1986).
37. K. D. Rothe, *Nucl. Phys. B269*, 269 (1986); R. Roskies and F. A. Schaposnik, *Phys. Rev. D 23*, 558 (1981); R. E. Gamboa Saravi, F. A. Schaposnik, and J. E. Solomin, *Nucl. Phys. B185*, 239 (1981); N. K. Nielsen, K. D. Rothe, and B. Schroer, *Nucl. Phys. B160*, 330 (1979).
38. A. M. Polyakov and P. B. Wiegmann, *Phys. Lett. 131B*, 121 (1983).
39. U. Lindström, N. K. Nielsen, M. Roček, and P. van Nieuwenhuizen, Stony Brook preprint.
40. M. Rocek, P. van Nieuwenhuizen, and C. Z. Zhang, *Ann. Phys.* (*N.Y.*) *172*, 348 (1986).
41. P. van Nieuwenhuizen, Proceedings of the Second Nobel Symposium on particle physics.
42. D. Nemeschansky and S. Yankielovicz, *Phys. Rev. Lett. 54*, 620 (1985).

Chapter 15

The Grassmann Oscillator*

Manuel Villasante

1. INTRODUCTION

I would like to present in this chapter a little toy. We start by considering the Lagrangian

$$L = \tfrac{1}{2}(\dot{q}^\alpha C_{\alpha\beta}\dot{q}^\beta - \omega^2 q^\alpha C_{\alpha\beta} q^\beta) \tag{1}$$

where the q^α are anticommuting variables, and the indices α and β run from 1 to n. The matrix $C_{\alpha\beta}$ has to be antisymmetric of course, and therefore the number n has to be even if we want the matrix $C_{\alpha\beta}$ to be invertible. This is the case we will consider.

I should mention here that Bryce De Witt in his book *Supermanifolds*, has also discussed the Fermi oscillator [2]. He considers, though, a Lagrangian with only one time derivative, in the anticommuting variables, and also his kinetic term is diagonal. The fact that we consider a kinetic term with two time derivatives should not be surprising any more, after the work by Stephen Adler (Chapter 1 of this volume).

* Work done in collaboration with R. J. Finkelstein [1].

MANUEL VILLASANTE • Department of Physics, University of California at Los Angeles, Los Angeles, California 90024.

Defining the symplectic adjoint by

$$\bar{q}_\alpha = q^\beta C_{\alpha\beta} \tag{2}$$

we can write the Lagrangian

$$L = \tfrac{1}{2}(\dot{\bar{q}}\dot{q} - \omega^2 \bar{q}q) \tag{3}$$

Now we can define the canonical momentum

$$p_\alpha = \frac{\partial L}{\partial \dot{q}^\alpha} \tag{4}$$

and also the Hamiltonian

$$H = \dot{q}^\alpha p_\alpha - L \tag{5}$$

In this way the Hamilton equations of motion are

$$\dot{q}^\alpha = (\partial H/\partial p_\alpha)_r, \qquad \dot{p}_\alpha = -(\partial H/\partial q^\alpha) \tag{6}$$

where the subindex r means the right derivative, which is different from the left derivative, since we are dealing with Grassmann variables. In our case we get

$$p_\alpha = C_{\alpha\beta}\dot{q}^\beta \Rightarrow \dot{q}^\alpha = (C^{-1})^{\alpha\beta} p_\beta \tag{7}$$

so the Hamiltonian becomes

$$H = \tfrac{1}{2}(-pC^{-1}p + \omega^2 qCq) \tag{8}$$

and the equations of motion are

$$\dot{q}^\alpha = (C^{-1}p)^\alpha, \qquad \dot{p}_\alpha = -\omega^2 (Cq)_\alpha \tag{9}$$

From these last two equations we derive

$$\ddot{q}^\alpha + \omega^2 q^\alpha = 0 \tag{10}$$

and the conservation of the Hamiltonian

$$\dot{H} = 0 \tag{11}$$

Bryce De Witt gets the same equation of motion in his case. In the quantum case we have now the anticommunication relations

$$\{q^\alpha, q^\beta\} = 0, \qquad \{p_\alpha, p_\beta\} = 0, \qquad \{q^\alpha, p_\beta\} = -i\hbar\delta^\alpha_\beta \tag{12}$$

with $p = -i\hbar\partial/\partial q$.

2. SPECTRUM AND STATES

In a similar way to the bosonic case we can define creation and annihilation operators

$$a_{\pm} = \frac{1}{\sqrt{2}}(\omega q \mp iC^{-1}p), \qquad a_{+}^{\dagger} = a_{-} \tag{13}$$

which satisfy the anticommunication relations

$$\{a_{-}^{\alpha}, a_{-}^{\beta}\} = 0 = \{a_{+}^{\alpha}, a_{+}^{\beta}\}, \qquad \{a_{-}^{\alpha}, \bar{a}_{\beta}^{+} = \omega\delta_{\beta}^{\alpha}\} \tag{14}$$

where

$$\bar{a}_{\beta}^{+} = a_{+}^{\alpha}C_{\alpha\beta}$$

In terms of these operators, the Hamiltonian is given by

$$\begin{aligned} H &= \tfrac{1}{2}(a_{-}Ca_{+} + a_{+}Ca_{-}) = \tfrac{1}{2}(\bar{a}^{+}a_{-} - a_{-}\bar{a}^{+}) \\ &= \bar{a}^{+}a_{-} - \tfrac{1}{2}n\omega = -a_{-}\bar{a}^{+} + \tfrac{1}{2}n\omega \end{aligned} \tag{15}$$

and therefore we obtain the commutators

$$[H, a^{\alpha}{}_{\pm}] = \pm\omega a_{\pm}^{\alpha} \tag{16}$$

If $|m\rangle$ is an eigenstate of H, with eigenvalue ε_m, we have

$$Ha_{\pm}^{\alpha}|m\rangle = (\varepsilon_m \pm \omega)a_{\pm}^{\alpha}|m\rangle \tag{17}$$

showing that $a_{+}(a_{-})$ is a raising (lowering) operator in the spectrum of H.

It is clear by now, since the a_{+} are anticommuting, that there is going to be a finite number of states.

Using a well-known identity in quantum mechanics, we see that (13) can be written in a more convenient form

$$\begin{aligned} a_{-}^{\alpha} &= \frac{1}{\sqrt{2}} e^{-(\omega/2)\bar{q}q} \frac{\partial}{\partial \bar{q}_{\alpha}} e^{(\omega/2)\bar{q}q} \\ a_{+}^{\alpha} &= \frac{-1}{\sqrt{2}} e^{(\omega/2)\bar{q}q} \frac{\partial}{\partial \bar{q}_{\alpha}} e^{-(\omega/2)\bar{q}q} \end{aligned} \tag{18}$$

from which we immediately obtain the lowest and highest state:

$$\begin{aligned} &\text{lowest state:} \qquad a_{-}e^{-(\omega/2)\bar{q}q} = 0 \\ &\text{highest state:} \qquad a_{+}e^{(\omega/2)\bar{q}q} = 0 \end{aligned} \tag{19}$$

The "energies" (eigenvalues of the Hamiltonian) of these states are

$$He^{\pm\omega\bar{q}q/2} = \pm\tfrac{1}{2}n\omega e^{\pm\omega\bar{q}q/2} \tag{20}$$

as can be seen immediately from (15). We now proceed to construct the complete set of eigenfunctions, starting from the Grassmann Gaussian

$$\psi_0 = e^{-(\omega/2)\bar{q}q} \tag{21}$$

by applying the creation operators a_+^α. The expression

$$\psi^{\alpha_1\ldots\alpha_m} = a_+^{\alpha_1}\ldots a_+^{\alpha_m}\psi_0 = \left(-\frac{1}{\sqrt{2}}\right)^m e^{(\omega/2)\bar{q}q}\frac{\partial}{\partial\bar{q}_{\alpha_1}}\cdots\frac{\partial}{\partial\bar{q}_{\alpha_m}}e^{-(\omega/2)\bar{q}q}\psi_0 \tag{22}$$

is an eigenfunction of the Hamiltonian

$$H\psi^{\alpha_1\ldots\alpha_m} = E_m\psi^{\alpha_1\ldots\alpha_m} \tag{23}$$

with eigenvalue

$$E_m = m\omega - \tfrac{1}{2}n\omega \tag{24}$$

and the degeneracy of this value is $\binom{n}{m}$. There are $n+1$ eigenvalues, some positive and some negative. They are symmetric and the zero eigenvalue is one of them. We can write these eigenfunctions in a more suggestive form:

$$\psi^{\alpha_1\ldots\alpha_m} = e^{-(\omega/2)\bar{q}q}H^{\alpha_1\ldots\alpha_m}(q) \tag{25}$$

where

$$H^{\alpha_1\ldots\alpha_m}(q) = \left(-\frac{1}{\sqrt{2}}\right)^m e^{\omega\bar{q}q}\frac{\partial}{\partial\bar{q}_{\alpha_1}}\cdots\frac{\partial}{\partial\bar{q}_{\alpha_m}}e^{-\omega\bar{q}q} \tag{26}$$

are what we call the Grassmann–Hermite polynomials. These polynomials have properties of the standard Hermite polynomials. They satisfy corresponding recursion relations and differential equations. They even have a generating function from which we can derive orthogonality properties [3]

$$\int(dq)e^{-\omega\bar{q}q}H^{\alpha_1\ldots\alpha_m}(q)\bar{H}_{\beta_1\ldots\beta_l}(q) = \delta_l^m B_m A_m \delta_{[\beta_1}^{[\alpha_1}\ldots\delta_{\beta_m]}^{\alpha_m]} \tag{27}$$

where

$$\begin{aligned} B_n &= (-2)^{n/2}\,\mathrm{Pf}\,C, \qquad A_m = \omega^m(-)^{(m/2)(m-1)} \\ \bar{H}_{\alpha_1\ldots\alpha_m}(q) &= H^{\beta_1\ldots\beta_m}(q)C_{\beta_1\alpha_1}\cdots C_{\beta_m\alpha_m} \end{aligned} \tag{28}$$

and $\delta_{[\beta_1}^{[\alpha_1}\ldots\delta_{\beta_m]}^{\alpha_m]}$ is the completely antisymmetric part of the product of the Krönecker deltas (using a notation that is now standard in the literature) since the Grassmann–Hermite polynomials are obviously antisymmetric in all their indices. The integral is, of course, a Berezin integral, and Pf C is the Pfaffian of C, which, since C is antisymmetric, is just the square root of the determinant of C and appears all over where we deal with these integrals.

3. SOLUTION TO TIME-DEPENDENT SCHRÖDINGER EQUATION

Clearly the stationary state solution with "energy" E_m to the time-dependent Schrödinger equation is

$$\psi^{\alpha_1 \ldots \alpha_m}(q, t) = e^{-iE_m t} u^{\alpha_1 \ldots \alpha_m}(q) \tag{29}$$

where $u^{\alpha_1 \ldots \alpha_m}(q)$ is the $\psi^{\alpha_1 \ldots \alpha_m}(q)$ of the previous section multiplied by the factor needed to normalize it.

We can consider also an arbitrary time-dependent state:

$$\psi(q, t) = \int G(q, t; q_0 t_0)\psi(q_0, t_0)(dq_0) \tag{30}$$

where the Green's function can be computed in a straightforward way by the formula

$$G(q, t; q_0, t_0) = e^{(i/2)n\omega(t-t_0)} \sum_{m=0}^{n} \frac{\lambda}{m!} \bar{u}_{\alpha_1 \ldots \alpha_m}(q) u^{\alpha_1 \ldots \alpha_m}(q_0) \tag{31}$$

with $\lambda = e^{-i\omega(t-t_0)}$. In order to get a more compact expression, we are going to use the Fourier transform of the Grassmann Gaussian, which is just the identity

$$e^{-\omega\bar{\Phi}\Phi} = k \int (d\theta) e^{2i\omega\bar{\Phi}\theta} e^{-\omega\bar{\theta}\theta} \tag{32}$$

with

$$k = \frac{1}{(2\omega)^{n/2} \operatorname{Pf} C}$$

For the Grassmann–Hermite polynomials we have

$$\begin{aligned} H^{\alpha_1 \ldots \alpha_m}(q) &= \left(-\frac{1}{\sqrt{2}}\right)^m k e^{\omega\bar{q}q} \frac{\partial}{\partial \bar{q}_{\alpha_1}} \cdots \frac{\partial}{\partial \bar{q}_{\alpha_m}} \int e^{2i\omega\bar{q}Q} e^{-\omega\bar{Q}Q}(dQ) \\ &= \left(-\frac{1}{\sqrt{2}}\right)^m k e^{\omega\bar{q}q} \int (2i\omega)Q^{\alpha_1} \ldots (2i\omega)Q^{\alpha_m} e^{-\omega\bar{Q}Q + 2i\omega\bar{q}Q}(dQ) \end{aligned} \tag{33}$$

Replacing this in (31) we get

$$\begin{aligned} G(q, t; q_0, t_0) &= \frac{e^{(i/2)n\omega(t-t_0)}}{(-2\omega)^{n/2} \operatorname{Pf} C} k^2 e^{(\omega/2)(\bar{q}q + \bar{q}_0 q_0)} \sum_{m=0}^{n} \frac{(-2\lambda\omega)^m}{m!} \\ &\quad \times \int (\bar{Q}Q')^m e^{-\omega(\bar{Q}Q - 2i\bar{q}Q) - \omega(\bar{Q}'Q' - 2i\bar{q}_0 Q')}(dQ')(dQ) \end{aligned} \tag{34}$$

We can see here that the sum in m becomes just an exponential, therefore the integrals in Q, Q' can be done by completing the squares, which is a lot simpler here than in the case Professor Adler discussed.

The result is

$$G(q, t; q_0, t_0) = \frac{i^{n/2}}{\text{Pf } C}\left[\frac{\sin \omega(t - t_0)}{\omega}\right]^{n/2} e^{iS} \tag{35}$$

where S is the classical action for a path going from position q_0 at time t_0, to a position q at time t,

$$S = \frac{1}{2}\frac{\omega}{\sin \omega(t - t_0)}[(\bar{q}q + \bar{q}_0 q_0)\cos \omega(t - t_0) - 2\bar{q}q_0] \tag{36}$$

whose form is entirely identical to the bosonic case. On the other hand, the amplitude is now inverted. In particular the factor $[\sin \omega(t - t_0)/\omega]^{n/2}$ appears upside down with respect to the bosonic case, giving us a limited amplitude, instead of an arbitrarily high one, as required by the exclusion principle.

From the above Green's function, we get the free particle one in the limit $\omega \to 0$

$$G(q, t; q_0, t_0) = \frac{i^{n/2}}{\text{Pf } C}(t - t_0)^{n/2} \exp\left[\frac{i}{2(t - t_0)}(\bar{q} - \bar{q}_0)(q - q_0)\right] \tag{37}$$

and both Green's functions give in the limit $t \to t_0$:

$$\lim_{t \to t_0} G(q, t; q_0, t_0) = \delta(q - q_0) \tag{38}$$

where $\delta(q - q_0)$ is the Berezin delta function. So this Green's function has the right limits.

After seeing all these results, I would like to stress the remarkable similarities between the bosonic and fermionic harmonic oscillators. In both cases we get the *same* functions. They may look different, given the differences between bosonic and fermionic variables. For instance, the Berezin delta function looks quite different from the Dirac delta function, but they play identical roles upon integration and they are the limits when $t \to t_0$ of the Green's function in the corresponding cases. Equally striking are the similarities between the stationary state wave functions. The similarities between the Hermite polynomials and the Grassmann-Hermite polynomials in their properties have already been mentioned even though the latter ones have explicit spinorial indices.

4. OPERATOR HAMILTON–JACOBI EQUATION

One can find the Green's function of the previous section using Schwinger's action principle. From this principle we get the expression for the Green's function:

$$G(q_2, t_2; q_1, t_1) = \langle q_2, t_2; q_1, t_1 \rangle = e^{(i/\hbar)\mathcal{W}'_{21}} \tag{39}$$

where $\mathcal{W}'_{21}$ is the matrix element of the time-ordered action which satisfies the operator Hamilton–Jacobi equation

$$H\left(\frac{\partial \mathcal{W}}{\partial q}, q\right) + \frac{\partial \mathcal{W}}{\partial t} = 0 \tag{40}$$

which, in our case becomes

$$-\frac{1}{2}\frac{\partial \mathcal{W}}{\partial q^\alpha}(C^{-1})^{\alpha\beta}\frac{\partial \mathcal{W}}{\partial q^\beta} + \frac{1}{2}\omega^2 q^\alpha C_{\alpha\beta} q^\beta + \frac{\partial \mathcal{W}}{\partial t} = 0 \tag{41}$$

In order to solve this equation, we try the same ansatz that works in the bosonic case [4]

$$\mathcal{W} = S_{\text{class}} + \Phi(\omega, t) = \frac{1}{2}\frac{\omega}{\sin \omega t}\{(\bar{q}q + \bar{q}_0 q_0)\cos \omega t - 2\bar{q}q_0\} + \Phi(\omega, t) \tag{42}$$

$\mathcal{W}$ has to be different from the classical action since q does not anticommute with q_0, because they correspond to different times. Space limitations do not permit showing how this works, but the details can be found in Ref. 1. The result obtained is exactly the same as before and it can also be obtained using Feynman's method of path integrals.

5. APPLICATIONS

One could ask, why did we ever bother to play with this toy? As a matter of fact, we found the corresponding spectrum and eigenfunctions while looking at the massive representations of the super Poincaré algebra in higher dimensions [3].

To start with, one could consider a Lie algebra

$$[X_A, X_B] = i f^C_{AB} X_C \tag{43}$$

By associating to the operator X_A the expression $\bar{a}X_A a$, where the X_A are appropriate numerical matrices, so that

$$[\bar{a}X_A a, \bar{a}X_B a] = i f^C_{AB}\bar{a}X_C a \tag{44}$$

and the a's satisfy anticommutation relations

$$\{\bar{a}_\alpha, a^\beta\} = \delta_\alpha^\beta$$

This works quite similarly to the bosonic case, once more. But nobody cares about plain Lie algebras anymore.

For the super Poincaré algebra in d dimensions, SP_d the little algebra is formed by the generators [3]

$$U_{AB} = -\frac{i}{4}\frac{P^E}{P^2}\bar{D}\Gamma_{EAB}D \tag{45}$$

where Γ_{EAB} is the totally antisymmetrized product of three Dirac matrices $\Gamma_{[E}\Gamma_A\Gamma_{B]}$.

The operators U_{AB} satisfy an $SO(d-1)$ algebra in the rest frame, and D are the so-called covariant derivatives of the algebra. The eigenfunctions of the quadratic Casimir satisfy

$$\bar{D}D\psi_m = \varepsilon_m\psi_m \tag{46}$$

and the spectrum is

$$\varepsilon_m = 0, \pm 2M, \pm 4M, \ldots, \pm 2kM \tag{47}$$

where M is the "mass" given by eigenvalue of P^2 and

$$k = 2^{[d/2]-2} \qquad (k = 8 \text{ in } 11 \text{ dimensions})$$

$\bar{D}D$ corresponds to the Hamiltonian and the raising and lowering operators, which change from one representation to another, are the positive and negative energy projections of D:

$$D_\pm = \Lambda_\pm D \qquad \text{with } \Lambda_\pm = \frac{1}{2M}(M \pm \not{P}) \tag{48}$$

The ψ_m in (46) are given by

$$\psi_m = e^{(M/4)\bar{\theta}\theta}\sum_{p=0}^{d/2} H^{\alpha_1\ldots\alpha_p\beta_1\ldots\beta_m}(\theta)\bar{F}_{\alpha_1\ldots\alpha_p;\beta_1\ldots\beta_m}(x) \tag{49}$$

where the Grassmann-Hermite polynomials are in this case

$$H^{\alpha_1\ldots\alpha_p}(\theta) = e^{-(M/4)\bar{\theta}\theta}\frac{\partial}{\partial\bar{\theta}_{\alpha_1}}\cdots\frac{\partial}{\partial\bar{\theta}_{\alpha_p}}e^{(M/4)\bar{\theta}\theta} \tag{50}$$

and the fields $\bar{F}_{\alpha_1\ldots\alpha_p;\beta_1\ldots\beta_m(x)}$ are antisymmetric Bargmann-Wigner multi-spinors obeying

$$\begin{aligned} \Lambda_{-\ \gamma}^{\alpha_s}\bar{F}_{\alpha_1\ldots\alpha_s\ldots\alpha_p;\beta_1\ldots\beta_m} = 0 \\ \Lambda_{+\ \gamma}^{\beta_s}\bar{F}_{\alpha_1\ldots\alpha_p;\beta_1\ldots\beta_s\ldots\beta_m} = 0 \end{aligned} \tag{51}$$

Further details can be found in Ref. 3. I will just say here that the massive representations of the super Poincaré algebra in any number of dimensions (meaning, of course, that Majorana spinors must exist: 5, 6, 7, mod 8 are excluded) are connected to the Grassmann oscillator in the way I showed with one important exception: the ten-dimensional one. The reason is rather technical and I will not explain it here, but is ultimately connected to the fact that in ten dimensions there are Majorana–Weyl spinors. If this case is also connected to the oscillator, it is going to be in a more complicated way.

DISCUSSION

P. VAN NIEUWENHUIZEN: I don't see why De Witt gets the same equation if he does not iterate it.

M. VILLASANTE: He does.

VAN NIEUWENHUIZEN: Then he introduces more solutions than he already has.

VILLASANTE: Yes.

R. JACKIW: Is a_+ the Hermitian conjugate of a_-?

VILLASANTE: Yes. The operator p is not Hermitian, it is anti-Hermitian. It is a constant matrix which must be imaginary so that the Lagrangian and the Hamiltonian are real (with the standard prescription $(q^\alpha q^\beta)^* = q^{*\beta} q^{*\alpha}$ for anticommuting numbers) and Hermitian.

M. HENNEAUX: Do you have negative norm states in your Hilbert space?

VILLASANTE: I haven't looked at that. One would have to define a norm which, if defined in the usual way, could give you a nilpotent number for instance.

VAN NIEUWENHUIZEN: You have shown here explicit construction of irreducible representations of the super-Poincaré algebra in terms of these creation and destruction operators. Now the representations are known already. They fall into three groups: massless ones and massive ones with and without central charges. Have you gotten representations for all three classes?

VILLASANTE: No, I was analyzing the representations where $P^2 = M^2$ is positive, without central charges.

VAN NIEUWENHUIZEN: One could extend that and I think it should also work in ten dimensions.

VILLASANTE: Yes, I think one could extend that. This procedure is "à la Sokatchev." It does not work in ten dimensions, for a very particular reason, which is related to the fact that in ten dimensions you can define Majorana–Weyl spinors.

C. ARAGONE: Then, the same problem appears in two dimensions.

VILLASANTE: Yes, but the procedure works in all dimensions where you can have Majorana spinors: 8, 9, 11, and 4.

REFERENCES

1. R. J. Finkelstein and M. Villasante, *Phys. Rev. D 33*(6), 1666 (1986).
2. B. S. De Witt, *Supermanifolds*, Cambridge University Press, Cambridge, 1984.
3. R. J. Finkelstein and M. Villasante, *J. Math. Phys. 27*, 1595 (1986).
4. R. J. Finkelstein, *Nonrelativistic Mechanics*, Benjamin, New York, 1973.

Index

www.ingramcontent.com/pod-product-compliance
Lightning Source LLC
LaVergne TN
LVHW052309200926
848481LV00002B/508

* 9 7 8 1 4 8 9 9 3 7 3 0 8 *